This book belongs to

Name: _____

www.math-knots.com

Cover Design by :
Gowri Vemuri

First Edition :
March , 2023

Author :
Gowri Vemuri

Edited by :
Raksha Pothapragada
Ritvik Pothapragada

Questions: mathknots.help@gmail.com

NOTE : CCSSO or NCTM is neither affiliated nor sponsors or endorses this product.

Dedication

This book is dedicated to:

My Mom, who is my best critic, guide and supporter.

To what I am today, and what I am going to become tomorrow,

is all because of your blessings, unconditional affection and support.

This book is dedicated to the

strongest women of my life,

my dearest mom

and

to all those moms in this universe.

G.V.

6

Math-Knots Grade level enrichment series covers all pre-K to Grade 10 common core math work books.

The Grade 6 math work book is aligned to common core curriculum and more , challenge level questions are included. Content is divided across weeks aligned to school calendar year.
Six assessments are provided based on the topics covered in the previous weeks. The assessments help them to identify learning gaps students can redo the previous weeks content to bridge their learning gaps as needed. End of the year assessment is provided online at the below URL.
www.a4ace.com

For more practice ,you can also subscribe at www.a4ace.com .
All practice sets, and assessments can be taken any number of times within the one-year subscription period.
All our content is written by industry experts with over 30 years of experience.
A4ace.com is part of Math-Knots LLC
Math-Knots is your one stop enrichment place. ~~Learn to think with us.

Note: Video explanations of content in the book is coming up.

Instructions to register on www.a4ace.com
1. Register as a parent and choose a category while registering.
2. Register the student and choose a category while registering.
3. You can add more categories from the dash board as needed.
4. After registering an automatic email will be sent for verification.
5. Verify the account by clicking on the link sent to your email.
6. Login with your credentials and navigate to the dash board.
7. Click on the free test and follow through the instructions.
8. Free test will navigate to a payment page and it will say $0.
    You just have to click Paypal button and you can take the test.
9. Any issues please reach out to mathknots.help@gmail.com

8

 **Algebra 1**

Index

<table>
<tr><th>WEEK NO</th><th>TOPIC</th><th>PAGE NO</th></tr>
<tr><td>Week #1</td><td>Variable and Numerical Expressions</td><td>29-39</td></tr>
<tr><td>Week #2</td><td>Simplifying Expressions</td><td>40-48</td></tr>
<tr><td>Week #3</td><td>Simplifying Expressions</td><td>49-59</td></tr>
<tr><td>Week #4</td><td>Linear Equations</td><td>60-68</td></tr>
<tr><td>Week #5</td><td>Linear Equations</td><td>69-77</td></tr>
<tr><td>Week #6</td><td>Linear Equations</td><td>78-90</td></tr>
<tr><td>Week #7</td><td>Absolute Value Equations</td><td>91-99</td></tr>
<tr><td>Week #8</td><td>Literal Equations</td><td>100-118</td></tr>
<tr><td>Week #9</td><td>ASSESSMENT 1</td><td>119-129</td></tr>
<tr><td>Week #10</td><td>Inequalities</td><td>130-139</td></tr>
<tr><td>Week #11</td><td>Inequalities</td><td>140-152</td></tr>
<tr><td>Week #12</td><td>Compound Inequalities</td><td>153-174</td></tr>
<tr><td>Week #13</td><td>Graphing Inequalities</td><td>175-220</td></tr>
<tr><td>Week #14</td><td>Functions</td><td>221-242</td></tr>
<tr><td>Week #15</td><td>Functions</td><td>243-276</td></tr>
<tr><td>WEEK#16</td><td>ASSESSMENT 2</td><td>277-292</td></tr>
<tr><td>Week #17</td><td>Slopes</td><td>293-315</td></tr>
<tr><td rowspan="2">Week #18</td><td>Slope Intercept form</td><td></td></tr>
<tr><td>Point Slope & Standard form of straight line</td><td>316-351</td></tr>
</table>

Answer Keys 352 - 381

PEDMAS (or PEMDAS) rules :

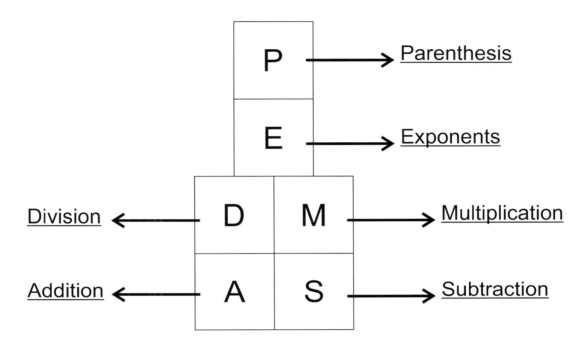

Steps to solve numerical or algebraic expressions and equations :

Step 1. Convert all division into a fraction (doing this way will avoid making silly mistakes).
Step 2. Solve the values within the parenthesis using PEDMAS rules.
   Within the parenthesis follow step 3 to step 8 as applicable.
Step 3. Check for anymore simplifications within the parenthesis using PEDMAS rules.
Step 4. Check for any exponential values if exists simplify otherwise proceed to step 5.
Step 5. Check for division and simplify as needed, if no division values exists proceed to step 6.
Step 6. Check for multiplications and simplify, if no multiplication exists proceed to step 7.
Step 7. Check for addition and simplify. Remember that a number should be considered along
   with its sign associated before the number. If no addition exists proceed to step 8.
Step 8. Check for subtraction and simplify. Remember that a number should be considered along
   with its sign associated before the number.

Rules of addition and subtraction of integers should be applied as needed for step 7 and 8.
Rules of multiplication and division of integers should be applied as needed in
step 1 through step 6.

           www.math-knots.com | www.a4ace.com

Order of operations :

Example 10 :

$(12 + 4)^2 \div 8 - 3^2 \times 2 + 5 \times 2 - 10$

Using PEDMAS rules we need to solve the numerical expressions.

Step 1 : (Convert ÷ to a fraction and rewrite the numerical expression)

$\dfrac{(12 + 4)^2}{8} - 3^2 \times 2 + 5 \times 2 - 10$

Step 2 : (First solve the parenthesis)

$\dfrac{(16)^2}{8} - 3^2 \times 2 + 5 \times 2 - 10$

Step 3 : Check for any more parenthesis to simplify if none are there then check for exponent values and simplify them.

$\dfrac{256}{8} - 9 \times 2 + 5 \times 2 - 10$

Step 4 : Multiply as needed in the numerical expression

$\dfrac{256}{8} - 18 + 10 - 10$

Step 5 : Divide as needed in the numerical expression

$32 - 18 + 10 - 10$

Step 6 : Simplify using addition and subtraction rules and pay attention to positive and negative signs.

$32 - 18 + 10 - 10 = 14 + 0 = 14$

12

Example 11 :

Simplify $x^2 + y(2x + y - 1) - 5$; where x = 5 and y = 3

Substitute the given values of x and y in the expression $x^2 + y(2x + y - 1) - 5$ and then solve.

Using PEDMAS rules we need to solve the numerical expressions.

Step 1 : (Convert ÷ to a fraction and rewrite the numerical expression if division operation is given in the expression otherwise proceed to step 2)

$5^2 + 3(2(5) + 3 - 1) - 5$

Step 2 : (First solve the parenthesis)

$5^2 + 3(2(5) + 3 - 1) - 5 = 5^2 + 3(10 + 3 - 1) - 5$

Step 3 : Check for any more parenthesis to simplify if none are there then check for exponent values and simplify them.

$5^2 + 3(10 + 3 - 1) - 5 = 5^2 + 3(13 - 1) - 5 = 5^2 + 3(12) - 5$

$5^2 + 3(12) - 5 = 25 + 3(12) - 5$

Step 4 : Multiply as needed in the numerical expression

$25 + 3(12) - 5 = 25 + 36 - 5$

Step 5 : Divide as needed in the numerical expression

$25 + 36 - 5$ (there are no division operation in this example)

Step 6 : Simplify using addition and subtraction rules and pay attention to positive and negative signs.

$25 + 36 - 5 = 61 - 5 = 56$

A Polynomial in standard form is always written with terms in sequential order according to their highest degree of exponents (higher exponents to lower exponents).

Like Terms :

Two or more terms are said to be alike if they have the same variable and the same degree. Coefficients of like terms are not necessarily be same.

An expression is in its simplest form when

1.   All like terms are combined.
2.   All parentheses are opened and simplified.

Like Terms can combined by adding or subtracting their coefficients (pay attention to the positive and negative signs of the coefficient and apply rules of adding integers)

Example 1 : -2x + 5x + 7 = 3x + 7

> Note : -2x and 5x are like terms and can be combined using rules of integers

Example 2 : -11y + 5 + 8y - 7 = -3y - 2

> Note : -11y and 8y are like terms and can be combined using rules of integers. 5 and -7 are like terms and can be combined using rules of integers

Example 3 : -12a - 5a + 8 - 3 = -17a + 5

> Note : -12a and -5a are like terms and can be combined using rules of integers. 8 and -3 are like terms and can be combined using rules of integers

Example 4 : -5b + 7 - 3b + 2a - a + 10 = -8b + a + 17

> Note : -5b and -3b are like terms and can be combined using rules of integers. 2a and -a are like terms and can be combined using rules of integers. 7 and 10 are like terms and can be combined using rules of integers.

Combining like terms on the opposite side of the equal sign :

When the like terms are on opposite sides, we have to combine like terms by using the inverse operation and by undoing the equation.

Example 5 : -2x + 5 = -7x

$$
\begin{array}{r}
-2x + 5 = -7x \\
+7x \qquad 7x \\
\hline
-2x + 7x + 5 = -7x + 7x \\
5x + 5 = 0
\end{array}
$$

Solving equations using the distributive property :

The number in front of the parentheses needs to be multiplied with every term within the parentheses. After the distribution and opening up the parentheses, combine like terms and solve.

Distributing with the negative sign :

Remember to apply the integer rules of positive and negative numbers while distributing.

$$
\begin{array}{l}
+ \text{ X } + = + \\
- \text{ X } - = + \\
- \text{ X } + = - \\
+ \text{ X } - = -
\end{array}
$$

Example 6 : (a) 2(5x + 7) = 2(5x) + 2(7) = 10x + 14
(b) -7(3a + 8) = (-7)(3a) + (-7)(8) = -21a + (-56) = -21a - 56
(c) 3(-5b - 2) = (3)(-5b) - (3)(2) = -15b - 6
(d) 6(-4a + 5) = (6)(-4a) + (6)(5) = -24a + 30
(e) 4(2a - 8) = (4)(2a) - (4)(8) = 8a - 32
(f) -5(a - 7) = (-5)(a) - (-5)(7) = -5a - (-35) = -5a + 35
(g) -9(-2a + 10) = (-9)(-2a) + (-9)(10) = 18a + (-90) = 18a - 90
(h) -8(-5a - 6) = (-8)(-5a) - (-8)(6) = 40a - (-48) = 40a + 40a
(i) -(a + 7) = -a - 7
(j) -(x - 5) = (-1)(x) - (-1)(5) = -x - (-5) = -x + 5
(k) -(-a - b) = (-1)(-a) - (-1)(b) = a - (-b) = a + b
(l) -(-a + 2b) = (-1)(-a) + (-1)(2b) = a + (-2b) = a - 2b

Example 7 : $2x + 3 = x + 7$

$2x + 3 = x + 7$
$\underline{-x \; - 3 \;\; -x \; -3}$
$x + 0 = 0 + 4$
$x = 4$

Inverse operation for addition is subtraction

Example 8 : $7x + 5 = -3x + 25$

$7x + 5 = -3x + 25$
$\underline{3x \; -5 \quad 3x \quad -5}$
$10x + 0 = 0 + 20$
$10x = 20$
$\dfrac{\cancel{10}x}{\cancel{10}} = \dfrac{\cancel{20}^2}{\cancel{10}}$

$\boxed{x = 2}$

Inverse operation for addition is subtraction and vice versa

Inverse operation for multiplication is division

Example 9 : $\dfrac{2x}{5} + 5 = 15$

$\dfrac{2x}{5} + 5 = 15$
$\underline{\qquad\quad -5 \quad -5}$
$\dfrac{2x}{5} + 0 = 10$

$\dfrac{2x}{5} = 10$

$5 \cdot \dfrac{2x}{\cancel{5}} = 5 \cdot 10$

$\dfrac{\cancel{2}x}{\cancel{2}} = \dfrac{\cancel{50}^{25}}{\cancel{2}}$

$\boxed{x = 25}$

Inverse operation for addition is subtraction and vice versa

Inverse operation for division is multiplication

Inverse operation for multiplication is division

www.math-knots.com | www.a4ace.com

Inequality :

An inequality is a relation between two expressions that are not equal. As a mathematical statement an inequality states one side of the equation is less than, less than or equal to or greater than or greater than equal to the other side.

If the inequality has **less than** or **greater than** symbol,
1.   The graph starts with the open circle.
2.   For less than the graphing line goes toward the left.
3.   For greater than the graphing line goes toward the right.

If the inequality has **less than or equal to** or **greater than or equal** to symbol,
1.   The graph starts with the closed circle.
2.   For less than or equal to the graphing line goes toward the left.
3.   For greater than or equal to the graphing line goes toward the right.

Inequality statement	Inequality verbal expression	Inequality graph
x > -3 or -3 < x	x is greater than -3	-4 -3 -2 -1 0 1 2 3
x < 3 or 3 > x	x is less than 3	-3 -2 -1 0 1 2 3 4
x >= -1 or -1 <= x	x is greater than or equal to -1	-3 -2 -1 0 1 2 3
x <= 1 or 1 <= x	x is less than or equal to 1	-3 -2 -1 0 1 2 3

Basic inequalities :

Solving inequalities is same as solving for an equation except for one special rule.

                   www.math-knots.com | www.a4ace.com

Example 10 : x + 9 > 11
Step 1 (subtract 9 from both sides) : x + 9 - 9 > 11 - 9
Step 2 (combine like terms) : x > 2

Example 11 : 2x + 5 > 10
Step 1 (subtract 5 from both sides) : 2x + 5 - 5 > 10 - 5
Step 2 (combine like terms) : 2x > 5
Step 3 (divide both sides by the coefficient of x which is 2) : $\dfrac{\overset{1}{\cancel{2}x}}{\underset{1}{\cancel{2}}} > \dfrac{5}{2}$

Step 4 (simplify both sides as needed) : x > $\dfrac{5}{2}$

Example 12 : 5x - 1 > 9
Step 1 (add 1 to both sides) : 5x - 1 + 1 > 9 + 1
Step 2 (combine like terms) : 5x > 10
Step 3 (divide both sides by the coefficient of x which is 5) : $\dfrac{\overset{1}{\cancel{5}x}}{\underset{1}{\cancel{5}}} > \dfrac{10}{5}$

Step 4 (simplify both sides as needed) : x > 2

Example 13 : 2x - 8 > -11
Step 1 (add 8 to both sides) : 2x - 8 + 8 > -11 + 8
Step 2 (combine like terms) : 2x > -3
Step 3 (divide both sides by the coefficient of x which is 2) : $\dfrac{\overset{1}{\cancel{2}x}}{\underset{1}{\cancel{2}}} > \dfrac{-3}{2}$

Step 4 (simplify both sides as needed) : x > $\dfrac{-3}{2}$

Example 14 : -4x + 7 > 10
Step 1 (subtract 7 from both sides) : -4x + 7 - 7 > 10 - 7
Step 2 (combine like terms) : -4x > 3
Step 3 (divide both sides by the coefficient of x which is -4) : $\dfrac{\overset{1}{\cancel{-4}x}}{\underset{1}{\cancel{-4}}} < \dfrac{3}{-4}$

> When an inequality is multiplied or divided with negative number, the inequality changes to the opposite

Step 4 (simplify both sides as needed) : x < $\dfrac{3}{-4}$

x < $\dfrac{-3}{4}$

www.math-knots.com | www.a4ace.com

Example 15 : -5x - 1 < -11
Step 1 (add 1 to both sides) : -5x - 1 + 1 < -11 + 1
Step 2 (combine like terms) : -5x < -10
Step 3 (divide both sides by the coefficient of x which is -5) : $\dfrac{\overset{1}{\cancel{-5}}x}{\underset{1}{\cancel{-5}}} > \dfrac{\overset{2}{\cancel{-10}}}{\underset{1}{\cancel{-5}}}$

> When an inequality is multiplied or divided with negative number, the inequality changes to the opposite

Step 4 (simplify both sides as needed) : x > 2

Example 16 : -8x + 3 < -30
Step 1 (subtract 3 from both sides) : -8x + 3 - 3 < -30 - 3
Step 2 (combine like terms) : -8x < -33
Step 3 (divide both sides by the coefficient of x which is -8) : $\dfrac{\overset{1}{\cancel{-8}}x}{\underset{1}{\cancel{-8}}} > \dfrac{-33}{-8}$

Step 4 (simplify both sides as needed) : x > $\dfrac{-33}{-8}$

$$x > \dfrac{33}{8}$$

$$x > 4\dfrac{1}{8}$$

Compound Inequalities :

x < 0 or x ≥ 5 means all values less than 0 or 5 and more. In other words we are excluding the values 0,1,2,3,4.

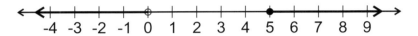

Absolute Value :

Absolute value of a number is its distance from 0. Since the distance cannot be negative absolute value is always positive.

$$|7| = 7 \qquad\qquad |-11| = 11$$

$$|2.3| = 2.3 \qquad\qquad |-0.75| = 0.75$$

Equations involving absolute values can be solved similar to regular algebraic equations solving. Absolute value should be treated as parentheses when applying PEDMAS rules.

Steps to solve absolute value equations.

Step 1 : Solve the expression within the absolute value.
(As applicable with PEDMAS rules)

Step 2 : Isolate the absolute value to one side of the equation.

Step 3 : Verify the value on the other side of the equation.
If the value is positive move to step 4.
If the value is negative there is no solution.

Step 4 : The expression inside the absolute value equals to positive and negative values of the other side of the equation.

Step 5 : Make the expression equal to positive value and solve for the variable.

Step 6 : Make the expression equal to negative value and solve for the variable.

Step 7 : The value obtained in step 5 and step 6 or the solutions to the absolute value equation.

Note : Absolute value equations can have two solutions. Since the absolute value of a number and its opposite are the same.
Absolute value can never be negative.

Absolute value Inequalities :

Absolute value inequalities are similar to absolute value equations.
Absolute value inequalities can have the below solutions
1. Two solutions
2. No solution
3. All real numbers

Steps to solve absolute value Inequalities are similar to solving the absolute value equations.

$|x| < a$ can be rewritten as $-a < x < a$ (where a is positive)
can also be written as $x < a$ **and** $x > -a$

$|x| \leq a$ can be rewritten as $-a \leq x \leq a$ (where a is positive)
can also be written as $x \leq a$ **and** $x \geq -a$

$|x| > a$ can be rewritten as $x > a$ **or** $x < -a$ (where a is positive)

Note : < or $\leq$ are represented by the word **and**
         > or $\geq$ are represented by the word **or**

SLOPE :

Slope of a straight line is how far the line is away from the horizontal line in other words how slanted or angular the straight line is.

Slope of a straight line is the rate of change for a given set of points.
Example : (x_1 , y_1) , (x_2 , y_2)

$$\text{Slope (m)} = \frac{\text{Rise}}{\text{Run}} = \frac{\text{difference in the y coordinates}}{\text{difference in the x coordinates}} = \frac{y_2 - y_1}{x_2 - x_1} = \frac{y_1 - y_2}{x_1 - x_2}$$

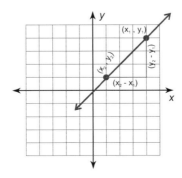

www.math-knots.com | www.a4ace.com

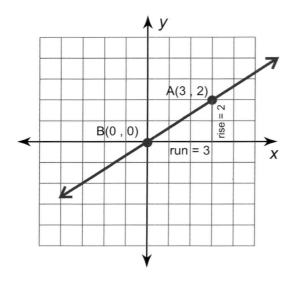

A straight line with a **positive** slope always **rises** from left to right

A(3 , 2) B(0 , 0)
↓   ↓   ↓   ↓
(x_2 , y_2) (x_1 , y_1)

Slope = $\dfrac{y_2 - y_1}{x_2 - x_1}$ = $\dfrac{2 - 0}{3 - 0}$ = $\dfrac{2}{3}$

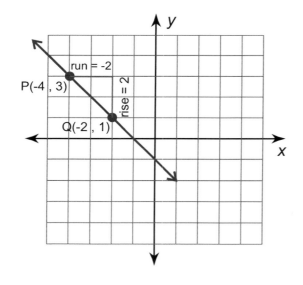

A straight line with a **negative** slope always **falls** from left to right

P(-4 , 3)   Q(-2 , 1)
↓   ↓   ↓   ↓
(x_2 , y_2) (x_1 , y_1)

Slope = $\dfrac{y_2 - y_1}{x_2 - x_1}$ = $\dfrac{3 - 1}{-4 - (-2)}$ = $\dfrac{2}{-4 + 2}$

= $\dfrac{2}{-2}$ = -1

www.math-knots.com | www.a4ace.com

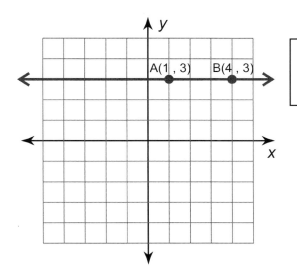

A straight line parallel to x axis always has a slope = 0

A(1 , 3)   B(4 , 3)
↓   ↓   ↓   ↓
(x_2 , y_2)   (x_1 , y_1)

Slope = $\dfrac{y_2 - y_1}{x_2 - x_1}$ = $\dfrac{3 - 3}{1 - 4}$ = $\dfrac{0}{-3}$ = 0

Note : A straight line of the form y = k, where k is a constant always has a zero slope.

Tip : When x coordinates are different and y coordinates are same slope is always zero.

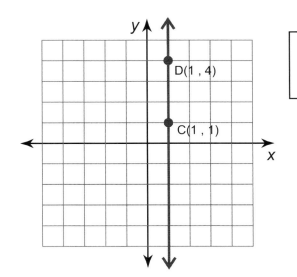

A straight line parallel to y axis always has an undefined slope

C(1 , 1)   D(1 , 4)
↓   ↓   ↓   ↓
(x_2 , y_2)   (x_1 , y_1)

Slope = $\dfrac{y_2 - y_1}{x_2 - x_1}$ = $\dfrac{1 - 4}{1 - 1}$ = $\dfrac{-3}{0}$ = undefined

Any number when you divide by 0 the value is undefined.

Note : A straight line of the form x = k, where k is a constant always has an undefined slope.

Tip : When x coordinates are same slope is undefined.

Slope intercept form of the straight line :

y = mx + b is the slope intercept form of the straight line where m is the slope of the straight line and b is the y intercept.

Examples : y = 7x - 1 ; Slope = 7 , y intercept = -1

y = -11x + 8 ; Slope = -11 , y intercept = 8

Standard form of the straight line :

ax + by = c is the Standard form of the straight line.

Example : 2x + 3y = 10

Point slope form of the straight line :

$y - y_1 = m(x - x_1)$ is the point slope form of the straight line where m is the slope and (x1 , y1) is any given point on the straight line.

Example : y - 2 = 3(x - 5) then slope of the straight line is 3 and (5 , 2) is a given point on the straight line.

Example : Given two points A(2 , 3) and B(5 , 8). Find the equation of a straight line in point slope form. Also express it in slope intercept form and standard form.

Step 1 : Find the slope

$$m = \frac{y_2 - y_1}{x_2 - x_1} = \frac{8 - 3}{5 - 2} = \frac{5}{3}$$

Step 2 : Substitute the slope value obtained in step 1 and any one point A or B in the equation $y - y_1 = m(x - x_1)$

$$y - 3 = \frac{5}{3} (x - 2)$$ Equation of the straight line in point slope form

Step 3 : Simplify the equation obtained in step 2 to rewrite in the form of y = mx + b

$$3 (y - 3) = \frac{5}{\cancel{3}} \cancel{3}.(x - 2)$$

                           www.math-knots.com | www.a4ace.com

Algorithm 1

Notes

$3y - 9 = 5x - 10$

$3y - 9 + 9 = 5x - 10 + 9$

$$\frac{3y}{3} = \frac{5x - 1}{3}$$

$$\boxed{y = \frac{5}{3}x - \frac{1}{3}}$$ Equation of the straight line in slope intercept form

Step 4 : Ax + By = C is the standard form of the straight line

Lets use the point slope of the straight line obtained in step 2.

$$y - 3 = \frac{5}{3}(x - 2)$$

$$3(y - 3) = \frac{5}{\cancel{3}}\cancel{3}.(x - 2)$$

$3y - 9 = 5x - 10$

$3y - 9 + 10 = 5x - 10 + 10$

$3y + 1 = 5x$

$3y + 1 - 3y = 5x - 3y$

$\boxed{5x - 3y = 1}$ Standard form of the straight line.

Important :
To find the x intercept of a given straight line substitute y = 0 and solve the equation. The value obtained is the x intercept.

Important :
To find the y intercept of a given straight line substitute x = 0 and solve the equation. The value obtained is the y intercept.

---

www.math-knots.com | www.a4ace.com

Graphing quadratic functions :

An equation of the standard form $y = ax^2 + bx + c$ is a quadratic function that has an x^2

The graph of a quadratic function is called as a parabola. The parabola is always symmetrical at its vertex. The vertex is a point where the graph changes its direction.

1. If a is positive in the quadratic function $y = ax^2 + bx + c$ then the parabola opens up.

2. If a is negative in the quadratic function $y = ax^2 + bx + c$ the parabola opens down.

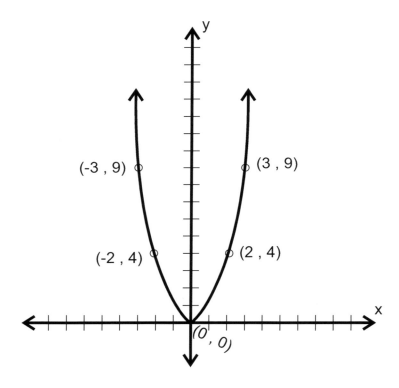

The graph of $y = x^2$ opens up as the coefficient of x^2
is positive and the vertex is (0,0) which is the lowest
point on the graph. The graph is symmetrical at the vertex.

 www.math-knots.com | www.a4ace.com

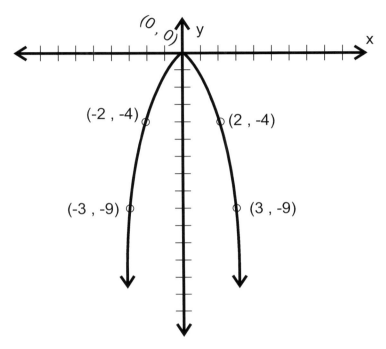

The graph of $y = -x^2$ opens down as the coefficient of x^2 is negative and the vertex is (0,0) which is the highest point on the graph. The graph is symmetrical at the vertex.

Finding x intercept :

The x intercept of any function are the values of x when y = 0.

1. For linear equations we can set y = 0 and find the value of x which is the x intercept.

2. For quadratic equations, factor the quadratic equation as explained in the previous sections and make it equal to 0 to find the values of x, which are x intercepts.

> x intercept is the point where the function value becomes 0.

Example : $y = 2x^2 - 9x - 5 = (x - 5)(2x + 1)$

To find x intercept make y = 0
so, $2x^2 - 9x - 5 = 0 = (x - 5)(2x + 1) = 0$

$(x - 5) = 0$, $(2x + 1) = 0$

$x = 5$ and $x = \dfrac{-1}{2}$

So the x intercepts are 5 and $\dfrac{-1}{2}$

If we plot a graph, the graph should touch at (5,0) and ($\dfrac{-1}{2}$, 0)

www.math-knots.com | www.a4ace.com

Finding y intercept :

The y intercept of any function are the values of y when x = 0.

1. For linear equations we can set x = 0 and find the value of y which is the y - intercept.

2. For quadratic equations, substitute x = 0 to find the values of y - intercept.

      y intercept is the point when x = 0.

Example : $y = 2x^2 - 9x - 5$

To find y intercept substitute x = 0
so, $2x^2 - 9x - 5 = y$
$y = 2(0^2) - 9(0) - 5$
$y = -5$

y - intercept is -5 for the quadratic equation $2x^2 - 9x - 5$.
The graph of the quadratic function $y = 2x^2 - 9x - 5$ touches the y axis at the point (0,-5)

Finding x intercept using quadratic formula :

The quadratic equation $y = ax^2 + bx + c = 0$ can also be factored by using the quadratic

formula $\dfrac{-b \pm \sqrt{b^2 - 4ac}}{2a}$

Step 1 : Make sure the quadratic equation is in the standard form $ax^2 + bx + c$ (convert it into standard form if it is not given)

Step 2 : If y is given then set y = 0.

Step 3 : Note down the values of "a" , "b" , "c". Make sure to consider the signs before each term.

Step 4 : Substitute the values "a" , "b" , "c" into the quadratic formula.

Step 5 : Since y = 0 the values found using the quadratic formula are the x - intercepts.

> It is not always possible to find the factors of the quadratic equation. In such a scenario the quadratic formula plays an important role. However we should get the same x - intercept values with both the methods.

1) Express the below as algebraic expression or equation or inequality.

the product of 6 and k

A) $k^6 < 26$ B) $6k$

C) $\dfrac{k}{2}$ D) $k - 6 > 21$

2) Express the below as algebraic expression or equation or inequality.

11 decreased by 9

A) $2 \cdot 11$ B) $11 - 9$

C) $11 + 9$ D) $\dfrac{9}{11}$

3) Express the below as algebraic expression or equation or inequality.

16 decreased by 12

A) $16 - 12$ B) $12 - 16$

C) $12 \cdot 16$ D) 16^{12}

4) Express the below as algebraic expression or equation or inequality.

25 decreased by 8

A) $25 - 8$ B) 25^3

C) 8^{25} D) $\dfrac{8}{25} > 34$

5) Express the below as algebraic expression or equation or inequality.

the product of 8 and 11

A) $\dfrac{8}{11} = 25$ B) $8 \cdot 11$

C) 11^8 D) 8^2

6) Express the below as algebraic expression or equation or inequality.

3 squared

A) $u \cdot 3$ B) $3 + u$

C) $\dfrac{3}{u}$ D) 3^2

7) Express the below as algebraic expression or equation or inequality.

26 decreased by m

A) $\dfrac{m}{26}$ B) $m - 26 = 5$

C) $\dfrac{26}{m}$ D) $26 - m$

8) Express the below as algebraic expression or equation or inequality.

the quotient of w and 8

A) $w \cdot 8$ B) $w^8 < 18$

C) $\dfrac{w}{8}$ D) $\dfrac{8}{w}$

9) Express the below as algebraic expression or equation or inequality.

13 less than z

A) $z^2 \le 43$ B) $13^3 < 38$

C) $z - 13$ D) $13 - z$

10) Express the below as algebraic expression or equation or inequality.

the 2nd power of 6

A) $2 - 6 \ge 32$ B) 2^6

C) 6^2 D) $6 - 2$

11) Express the below as algebraic expression or equation or inequality.

the quotient of 16 and n

A) $16 + n$
B) $n \cdot 16$

C) $\dfrac{16}{n}$
D) 16^3

12) Express the below as algebraic expression or equation or inequality.

22 decreased by 17

A) $22 - 17$
B) $17 + 22$

C) $22 \cdot 17$
D) $17 - 22$

13) Express the below as algebraic expression or equation or inequality.

7 less than 19

A) $19 - 7$
B) $19 + 7$

C) $\dfrac{19}{2}$
D) $7 \cdot 19$

14) Express the below as algebraic expression or equation or inequality.

v decreased by 15

A) $15 - v < 42$
B) 15^{v}

C) $15 - v$
D) $v - 15$

15) Express the below as algebraic expression or equation or inequality.

the quotient of n and 6

A) $n - 6$
B) $6 + n$

C) $\dfrac{n}{6}$
D) n^3

16) Express the below as algebraic expression or equation or inequality.

the product of v and 8

A) $8 - v$
B) $v \cdot 8$

C) v^8
D) $v - 8$

17) Express the below as algebraic expression or equation or inequality.

the x power of 5

A) $x - 5$
B) $\dfrac{5}{x}$

C) 5^x
D) $\dfrac{x}{5}$

18) Express the below as algebraic expression or equation or inequality.

the sum of 6 and 9

A) 6^9
B) $6 + 9$

C) $\dfrac{6}{9}$
D) $\dfrac{9}{2} < 34$

19) Express the below as algebraic expression or equation or inequality.

10 increased by 9

A) $10 + 9$
B) $10 - 9 < 45$

C) 10^3
D) 10^2

20) Express the below as algebraic expression or equation or inequality.

the quotient of 33 and 3

A) 3^2
B) $\dfrac{3}{33} > 34$

C) $\dfrac{33}{3}$
D) $33 \cdot 3$

www.math-knots.com | www.a4ace.com

21) Express the below as algebraic expression or equation or inequality.

n less than 15

A) $n + 15$ B) $15 - n$

C) $15^2 \le 20$ D) 15^n

22) Express the below as algebraic expression or equation or inequality.

8 increased by y

A) $y^2 < 18$ B) $y - 8$

C) $8 - y$ D) $8 + y$

23) Express the below as algebraic expression or equation or inequality.

5 less than a

A) $\dfrac{5}{a}$ B) $a - 5$

C) $5 - a$ D) a^5

24) Express the below as algebraic expression or equation or inequality.

the difference of 10 and 3

A) $10 - 3$ B) $3 - 10$

C) $3 - 10 \le 30$ D) $\dfrac{10}{3}$

25) Express the below as algebraic expression or equation or inequality.

11 more than 10

A) 10^2 B) $10 - 11$

C) $10 + 11$ D) $11 - 10$

26) Express the below as algebraic expression or equation or inequality.

the 2nd power of n

A) n^2 B) $n - 2$

C) $2^n \ge 41$ D) $2 + n$

27) Write each as a verbal expression or equation or inequality.

$p \cdot 5$

A) the product of p and 5
B) p squared
C) the sum of p and 5
D) 5 more than p

28) Write each as a verbal expression or equation or inequality.

$\dfrac{r}{5} = 18$

A) 5 times r is equal to 18
B) the quotient of r and 5 is equal to 18
C) the product of 5 and r is equal to 18
D) the sum of 5 and r is equal to 18

29) Write each as a verbal expression or equation or inequality.

$r - 17 = 32$

A) the difference of 17 and r is 32
B) r cubed is 32
C) 17 times r is 32
D) the difference of r and 17 is 32

30) Write each as a verbal expression or equation or inequality.

$n^2 = 6$

A) the difference of 2 and n is 6
B) n squared is 6
C) the quotient of n and 2 is 6
D) the sum of n and 2 is 6

31) Write each as a verbal expression or equation or inequality.

$$p + 10 = 30$$

A) 10 less than p is 30
B) 10 cubed is 30
C) 10 more than p is 30
D) p cubed is 30

32) Write each as a verbal expression or equation or inequality.

$$n - 3 = 8$$

A) 3 times n is 8
B) n cubed is 8
C) the difference of 3 and n is 8
D) the difference of n and 3 is 8

33) Write each as a verbal expression or equation or inequality.

$$n - 5 < 20$$

A) 5 more than n is less than 20
B) 5 less than n is less than 20
C) n increased by 5 is less than 20
D) n less than 5 is less than 20

34) Write each as a verbal expression or equation or inequality.

$$10^2$$

A) 10 squared
B) 2 minus 10
C) the quotient of 2 and 10
D) the difference of 10 and 2

35) Write each as a verbal expression or equation or inequality.

$$c - 16 = 35$$

A) c less than 16 is 35
B) 16 less than c is 35
C) 16 plus c is 35
D) the sum of 16 and c is 35

36) Write each as a verbal expression or equation or inequality.

$$z - 5 < 34$$

A) 5 plus z is less than 34
B) the difference of z and 5 is less than 34
C) the difference of 5 and z is less than 34
D) half of z is less than 34

37) Write each as a verbal expression or equation or inequality.

$$b - 10$$

A) the product of b and 10
B) b to the 10th
C) the difference of 10 and b
D) the difference of b and 10

38) Write each as a verbal expression or equation or inequality.

$$25^2$$

A) n more than 25
B) the product of n and 25
C) n squared
D) 25 squared

39) Write each as a verbal expression or equation or inequality.

$$z - 20 = 21$$

A) 20 less than z is 21
B) the product of z and 20 is 21
C) z plus 20 is 21
D) z less than 20 is 21

40) Write each as a verbal expression or equation or inequality.

$$\frac{32}{4}$$

A) 4 to the 32th
B) the difference of 32 and 4
C) 32 to the 4th
D) the quotient of 32 and 4

www.math-knots.com | www.a4ace.com

41) Write each as a verbal expression or equation or inequality.

$$p + 8 = 18$$

A) 8 times p is 18
B) p increased by 8 is 18
C) 8 less than p is 18
D) the difference of 8 and p is 18

42) Write each as a verbal expression or equation or inequality.

$$\frac{n}{8} < 45$$

A) 8 less than n is less than 45
B) 8 minus n is less than 45
C) the quotient of 8 and n is less than 45
D) the quotient of n and 8 is less than 45

43) Write each as a verbal expression or equation or inequality.

$$n - 9 > 40$$

A) twice 9 is greater than 40
B) 9 less than n is greater than 40
C) 9 increased by n is greater than 40
D) the product of 9 and n is greater than 40

44) Write each as a verbal expression or equation or inequality.

$$n + 5 = 46$$

A) n cubed is 46
B) 5 more than n is 46
C) 5 minus n is 46
D) 5 less than n is 46

45) Write each as a verbal expression or equation or inequality.

$$n + 12$$

A) the 12th power of n
B) the sum of n and 12
C) n to the 12th
D) 12 cubed

46) Write each as a verbal expression or equation or inequality.

$$n + 8 < 48$$

A) the n power of 8 is less than 48
B) n decreased by 8 is less than 48
C) 8 more than n is less than 48
D) 8 decreased by n is less than 48

47) Write each as a verbal expression or equation or inequality.

$$x + 5 = 8$$

A) the sum of x and 5 is equal to 8
B) x less than 5 is equal to 8
C) the 5th power of x is equal to 8
D) x minus 5 is equal to 8

48) Write each as a verbal expression or equation or inequality.

$$\frac{x}{3}$$

A) the quotient of 3 and x
B) the quotient of x and 3
C) x decreased by 3
D) x times 3

49) Write each as a verbal expression or equation or inequality.

$$n + 6$$

A) n increased by 6
B) 6 squared
C) 6 cubed
D) twice n

50) Write each as a verbal expression or equation or inequality.

$$k + 5$$

A) k increased by 5
B) the 5th power of k
C) twice 5
D) k less than 5

51) Write each as a verbal expression or equation or inequality.

$$11 + x$$

A) half of 11

B) 11 increased by x

C) x squared

D) 11 to the x

52) Simplify the below expression.

$$((28 - 12 - 8)(2)) \div 2$$

A) 8   B) 18

C) 15   D) 12

53) Simplify the below expression.

$$20 + \left(10^2\right)\left(7 - 6\right) - 4$$

A) 116   B) 112

C) 123   D) 100

54) Simplify the below expression.

$$((35)(2)) \div (20 - 19 + 7 + 6)$$

A) 10   B) 14

C) 20   D) 5

55) Simplify the below expression.

$$(8 - 20 \div ((4)(5) - 10))(12)$$

A) 66   B) 77

C) 72   D) 85

56) Simplify the below expression.

$$((7)(2)) \div (6 + 2 - (18 - 17))$$

A) 19   B) 14

C) 2   D) 12

57) Simplify the below expression.

$$((30)(2)) \div (18 - 1 + 3)$$

A) 13   B) 3

C) 18   D) 21

58) Simplify the below expression.

$$15 - 11 - (4 - 2 - (20 - 19))$$

A) 10   B) 15

C) 23   D) 3

59) Simplify the below expression.

$$(10 - 2 + 9 + 47) \div (18 - 2)$$

A) 3   B) 4

C) 18   D) 9

60) Simplify the below expression.

$$20 \div ((2)(2 - 1)) + 18 - 3$$

A) 25   B) 40

C) 15   D) 7

61) Simplify the below expression.

(9 + (10) (2) + 11) ÷ 10 + 5

A) 9   B) 28

C) 6   D) 25

62) Simplify the below expression.

(17) (1 + 2 + 9) - (18 + 9)

A) 191   B) 166

C) 177   D) 161

63) Simplify the below expression.

60 ÷ (18 - 8 + 17 - 17) + 12

A) 0   B) 30

C) 18   D) 5

64) Simplify the below expression.

(5) (17) (3) - (9) (6 + 15)

A) 70   B) 82

C) 57   D) 66

65) Simplify the below expression.

$((2)(6))^2 + ((15)(2)) \div 5$

A) 150   B) 142

C) 137   D) 154

66) Simplify the below expression.

12 - ((1 + 2) ÷ (18 - 15) + 8)

A) 20   B) 10

C) 5   D) 3

67) Simplify the below expression.

17 + 6 - 42 ÷ ((6) (2) - 9)

A) 9   B) 15

C) 3   D) 14

68) Simplify the below expression.

(11) (((44 + 1) (2)) ÷ 9) + 11

A) 112   B) 121

C) 107   D) 130

69) Simplify the below expression.

((48) (2)) ÷ ((4) (12 - 6) - 18)

A) 20   B) 16

C) 19   D) 28

70) Simplify the below expression.

(39 + 8 - 19) ÷ 14 - (2 - 1)

A) 1   B) 20

C) 19   D) 6

71) Simplify the below expression.

(45 + 21 - 20 - 20) ÷ (1 + 12)

A) 8 B) 17

C) 9 D) 2

72) Simplify the below expression.

((15) (3) + 27 - 16) ÷ (8 - 4)

A) 1 B) 28

C) 14 D) 18

73) Simplify the below expression.

((42 - 9 - 3) (2)) ÷ 6

A) 24 B) 31

C) 12 D) 15

74) Simplify the below expression.

18 + 17 - (12 - 7) + 10 - 8

A) 23 B) 32

C) 14 D) 40

75) Simplify the below expression.

((13) (2) + 1 + 3 - 2) ÷ 14

A) 2 B) 4

C) 19 D) 7

76) Simplify the below expression.

(17 - 50 ÷ 5) (12) - (2) (13)

A) 50 B) 59

C) 39 D) 58

77) Simplify the below expression.

(9 - 7) (19 - (6 + 8 ÷ 8))

A) 11 B) 26

C) 24 D) 6

78) Evaluate the below expression, with the values given

$y ÷ 2 + y - z - 3 ÷ 3$

Where $y = 2$, and $z = 1$

A) 12 B) 0

C) 20 D) 1

79) Evaluate the below expression, with the values given

$p - (r - (r - (q - 10) ÷ 3))$

Where $p = 17$, $q = 13$, and $r = 8$

A) 30 B) 13

C) 15 D) 16

80) Evaluate the below expression, with the values given

$x y + x + z - y x$

Where $x = 15$, $y = 12$, and $z = 13$

A) 28 B) 17

C) 12 D) 39

www.math-knots.com | www.a4ace.com

81) Evaluate the below expression, with the values given

$$x - (y - y)(y + zx)$$

Where $x = 19$, $y = 19$, and $z = 20$

A) 19 B) 20

C) 36 D) 10

82) Evaluate the below expression, with the values given

$$x \div 4 + x - y + 20\,y$$

Where $x = 20$, and $y = 3$

A) 66 B) 85

C) 82 D) 91

83) Evaluate the below expression, with the values given

$$18 - x + 12\,z + y + z$$

Where $x = 1$, $y = 3$, and $z = 6$

A) 99 B) 98

C) 111 D) 90

84) Evaluate the below expression, with the values given

$$a - (ab - (b + b)) \div 4$$

Where $a = 18$, and $b = 4$

A) 2 B) 14

C) 3 D) 17

85) Evaluate the below expression, with the values given

$$q - p \div 3 - (q - q) \div 6$$

Where $p = 9$, and $q = 13$

A) 14 B) 12

C) 10 D) 18

86) Evaluate the below expression, with the values given

$$(7)\left(h^2 - \left(j - j \div 6\right)\right)$$

Where $h = 6$, and $j = 12$

A) 172 B) 195

C) 162 D) 182

87) Evaluate the below expression, with the values given

$$pr - (r + r - q \div 4)$$

Where $p = 19$, $q = 8$, and $r = 5$

A) 87 B) 73

C) 81 D) 99

88) Evaluate the below expression, with the values given

$$3 + z - z + x + 2y$$

Where $x = 10$, $y = 9$, and $z = 12$

A) 35 B) 17

C) 45 D) 31

89) Evaluate the below expression, with the values given

$$c + a + b - (c)(c - c)$$

Where $a = 6$, $b = 19$, and $c = 19$

A) 44 B) 64

C) 26 D) 48

90) Evaluate the below expression, with the values given

$$z^2 + z - \left(y + y\right) \div 2$$

Where $y = 11$, and $z = 14$

A) 200 B) 199

C) 205 D) 188

Algebra 1

Vol 1
Week 1
Verbal Expressions
& Equations

91) Evaluate the below expression, with the values given

$$m - (3 + p \div 2 + p \div 2)$$

Where $m = 19$, and $p = 10$

A) 12   B) 4

C) 10   D) 6

92) Evaluate the below expression, with the values given

$$j + (j - (h - j)^2) \div 5$$

Where $h = 17$, and $j = 14$

A) 21   B) 30

C) 14   D) 15

93) Evaluate the below expression, with the values given

$$m - (p + 11 + m + p) \div 6 ;$$

Where $m = 15$, and $p = 5$

A) 7   B) 9

C) 17   D) 3

94) Evaluate the below expression, with the values given

$$13 + (z)(y \div 5 + z) - y$$

Where $y = 5$, and $z = 13$

A) 190   B) 170

C) 205   D) 191

95) Evaluate the below expression, with the values given

$$(z + x)(x + y) + 12 - y$$

Where $x = 1$, $y = 4$, and $z = 1$

A) 38   B) 7

C) 26   D) 18

96) Evaluate the below expression, with the values given

$$(5)(z + x) - (9 - (6 - x))$$

Where $x = 5$, and $z = 15$

A) 99   B) 97

C) 92   D) 87

97) Evaluate the below expression, with the values given

$$11n - (18 - (n - n)) - p$$

Where $n = 20$, and $p = 14$

A) 188   B) 172

C) 196   D) 170

98) Evaluate the below expression, with the values given

$$1 + (a)(b - c) - (a - b)$$

Where $a = 16$, $b = 10$, and $c = 6$

A) 59   B) 55

C) 68   D) 66

99) Evaluate the below expression, with the values given

$$y - x + 13z - z^2$$

Where $x = 5$, $y = 13$, and $z = 3$

A) 47   B) 38

C) 23   D) 20

100) Evaluate the below expression, with the values given

$$j - 1 - (7 - (k)(k - k))$$

Where $j = 11$, and $k = 14$

A) 16   B) 2

C) 3   D) 6

www.math-knots.com | www.a4ace.com

101) Evaluate the below expression, with the values given

$$b - (c + (a - a)^2 \div 6)$$

Where $a = 17$, $b = 16$, and $c = 12$

A) 5   B) 21

C) 9   D) 4

102) Evaluate the below expression, with the values given

$$12 - z \div 6 + y + z + 14$$

Where $y = 8$, and $z = 6$

A) 27   B) 21

C) 24   D) 39

103) Evaluate the below expression, with the values given

$$r - (p - (15 - (p) (r - 8)))$$

Where $p = 6$, and $r = 10$

A) 18   B) 21

C) 9   D) 7

 **Algebra 1**

1) Which expression is equivalent to

9 x - 2 x

2) Which expression is equivalent to

14 - 5n + 20 - 2n

3) Which expression is equivalent to

4n + 16n

4) Which expression is equivalent to

4b - 11b

5) Which expression is equivalent to

v - 19 + 9v + 8

6) Which expression is equivalent to

16r - 18 + 10 + 6r

7) Which expression is equivalent to

-9a + 19a

8) Which expression is equivalent to

1 - 6n + 9 - 18n

9) Which expression is equivalent to

-20a - 7a

10) Which expression is equivalent to

2 + 16v + 20 + 9v

www.math-knots.com | www.a4ace.com

11) Which expression is equivalent to

15v - 17v

12) Which expression is equivalent to

17k - 14k

13) Which expression is equivalent to

19 + 14x + 17

14) Which expression is equivalent to

-15 + 19n + n - 15

15) Which expression is equivalent to

2x + 13x

16) Which expression is equivalent to

-3 + 20v + v + 1

17) Which expression is equivalent to

11n + 19n

18) Which expression is equivalent to

-6a + 2a

19) Which expression is equivalent to

-5n + 19n

20) Which expression is equivalent to

19x - 10 + 10x

   www.math-knots.com | www.a4ace.com

21) Which expression is equivalent to

6x + 3x

22) Which expression is equivalent to

17x - 20x

23) Which expression is equivalent to

-11x + 7x

24) Which expression is equivalent to

-12n - 18n

25) Which expression is equivalent to

15r + 5 + 1 - 14r

26) Which expression is equivalent to

-3 - 11a + 5a - 9

27) Which expression is equivalent to

2.2n - 6.3 + 4.4n + 2.7

28) Which expression is equivalent to

m - 6.3 - 0.581

29) Which expression is equivalent to

1.2m - 2.4m

30) Which expression is equivalent to

8.4r + 4.2 - 8.3r

42   www.math-knots.com | www.a4ace.com

31) Which expression is equivalent to

1 + 0.1p - 9.8

32) Which expression is equivalent to

2.2n - 6.3 - 2.31

33) Which expression is equivalent to

-7v + 8v - 7.95 - 3

34) Which expression is equivalent to

-7.6x - 2x

35) Which expression is equivalent to

-4.6x + 7.4x

36) Which expression is equivalent to

1 + 9.7m + 7m

37) Which expression is equivalent to

1.4k + 9.6k

38) Which expression is equivalent to

8.4 p - 1.9 p

39) Which expression is equivalent to

4.889 - 3x + 7.5 - 9x

40) Which expression is equivalent to

-9.6 m - 10 m

41)   Which expression is equivalent to

3.71 n - 6.1 n

42)   Which expression is equivalent to

-8.7 x + 2.8 x

43)   Which expression is equivalent to

-5.7 x - 1.7 x

44)   Which expression is equivalent to

-8.9 k - 9.6 k

45)   Which expression is equivalent to

-2.8 n - 7.6 n

46)   Which expression is equivalent to

3.4 - 0.5 k - 9.2 k

47)   Which expression is equivalent to

-8.3 n + 2.7 n

48)   Which expression is equivalent to

-5.9 n + 7.2 + 8.9 n

49)   Which expression is equivalent to

1.4 x + 7.3 x

50)   Which expression is equivalent to

-6.4 p + 2.5 p

51) Which expression is equivalent to

p - 0.4 - 6

52) Which expression is equivalent to

-8.7 x + 4.7 x

53) Which expression is equivalent to

9 (n - 7)

54) Which expression is equivalent to

7 (r - 1)

55) Which expression is equivalent to

6 (8 n - 10)

56) Which expression is equivalent to

2 (n - 8)

57) Which expression is equivalent to

-10 (4r - 4)

58) Which expression is equivalent to

2 (b - 9)

59) Which expression is equivalent to

5 (2 + 6 k)

60) Which expression is equivalent to

3 (1 - 9 x)

61) Which expression is equivalent to

$$-9 (1 + 8 n)$$

62) Which expression is equivalent to

$$- (1 + 6 x)$$

63) Which expression is equivalent to

$$9 (8 n - 8)$$

64) Which expression is equivalent to

$$2 (3 k - 3)$$

65) Which expression is equivalent to

$$10 (3 + 2 x)$$

66) Which expression is equivalent to

$$10 (n + 10)$$

67) Which expression is equivalent to

$$-5 (1 - 7 x)$$

68) Which expression is equivalent to

$$5 (1 - m)$$

69) Which expression is equivalent to

$$4 (- 7 x - 1)$$

70) Which expression is equivalent to

$$- 2 (1 + 4 x)$$

www.math-knots.com | www.a4ace.com

71) Which expression is equivalent to

$$9 (- 10 k - 3)$$

72) Which expression is equivalent to

$$3 (- 10 - 3 x)$$

73) Which expression is equivalent to

$$9 (2 n + 7)$$

74) Which expression is equivalent to

$$- 8 (- 4 k + 8)$$

75) Which expression is equivalent to

$$- 10 (r - 10)$$

76) Which expression is equivalent to

$$- 3 (3 m - 2)$$

77) Which expression is equivalent to

$$9 (10 m - 2)$$

78) Which expression is equivalent to

$$- 3 (x + 3)$$

79) Which expression is equivalent to

$$10 (x + 6)$$

80) Which expression is equivalent to

$$- 3.33 (x + 7.4)$$

81) Which expression is equivalent to

$$6.1 (1 - 1.5 n)$$

82) Which expression is equivalent to

$$- 4.79 (n - 2.3)$$

83) Which expression is equivalent to

$$- 9.5 (7.95 b - 5.1)$$

84) Which expression is equivalent to

$$0.7 (n - 3.6)$$

85) Which expression is equivalent to

$$- 1.05 (6.5 x + 9.1)$$

86) Which expression is equivalent to

$$- 1.6 (3.3 n - 8.3)$$

87) Which expression is equivalent to

$$7.6 (1 - 1.5 b)$$

1) Which expression is equivalent to

- 6.8 (1.7 - 6 v)

2) Which expression is equivalent to

3.4 (- 3.3 + 0.2 k)

3) Which expression is equivalent to

0.4 (1 - 4.2 m)

4) Which expression is equivalent to

2.3 (1 + 7.7 x)

5) Which expression is equivalent to

- 1.9 (9.8 - 4.94 m)

6) Which expression is equivalent to

- 1.6 (v + 7.6)

7) Which expression is equivalent to

- 8.3 (- 7.95 a + 5.1)

8) Which expression is equivalent to

- 9.6 (x + 1.2)

9) Which expression is equivalent to

5.3 (n - 2.42)

10) Which expression is equivalent to

- 2.3 (v + 3.7)

11) Which expression is equivalent to

- 6.1 (b - 1.6)

12) Which expression is equivalent to

- 0.6 (x + 6.06)

13) Which expression is equivalent to

3.9 (6.5 - 8.3 b)

14) Which expression is equivalent to

- 9.2 (4.5 n + 3.7)

15) Which expression is equivalent to

9.53 (-4.5 x + 4.9)

16) Which expression is equivalent to

- 10 (n - 0.3)

17) Which expression is equivalent to

- 5.2 (6.9 m + 0.9)

18) Which expression is equivalent to

- 7.4 (- 6.6 + 9.8 x)

19) Which expression is equivalent to

9.5 (3.6 a - 8.6)

20) Which expression is equivalent to

9.6 (- 4.6 - 8.9 n)

21) Which expression is equivalent to

- 7 + 6 (6a - 4)

22) Which expression is equivalent to

3 (n + 8) + 4 n

23) Which expression is equivalent to

8 (9 r - 10) - 4

24) Which expression is equivalent to

10 (7 n - 11) - 7 n

25) Which expression is equivalent to

3 + 5(1 + 11 x)

26) Which expression is equivalent to

- 5 n + 9 (n + 3)

27) Which expression is equivalent to

2x - 5 (5 x + 1)

28) Which expression is equivalent to

5 (5 n + 12) + 7 n

29) Which expression is equivalent to

- 2 n + 11 (8 + 6 n)

30) Which expression is equivalent to

10 + 11 (7 n - 7)

31) Which expression is equivalent to

$$-6v + 6(v - 7)$$

32) Which expression is equivalent to

$$12(1 - 6b) + 6$$

33) Which expression is equivalent to

$$9x - 11(x + 3)$$

34) Which expression is equivalent to

$$-8(5 + 11x) + 3$$

35) Which expression is equivalent to

$$-9k + 5(8k + 4)$$

36) Which expression is equivalent to

$$-8n - 3(n + 11)$$

37) Which expression is equivalent to

$$-12(n - 1) - 3n$$

38) Which expression is equivalent to

$$-9(m + 12) - 3m$$

39) Which expression is equivalent to

$$-a - (12a - 12)$$

40) Which expression is equivalent to

$$6(9 - p) + 4p$$

41) Which expression is equivalent to

$$-7(-4r-10)-8$$

42) Which expression is equivalent to

$$-8x-12(3x-10)$$

43) Which expression is equivalent to

$$2(x-7)-11$$

44) Which expression is equivalent to

$$11n-11(7-10n)$$

45) Which expression is equivalent to

$$-(k-8)+5k$$

46) Which expression is equivalent to

$$-6(-12r-9)-r$$

47) Which expression is equivalent to

$$-(-9-11x)-3(-3-12x)$$

48) Which expression is equivalent to

$$3(r-7)+4(r+11)$$

49) Which expression is equivalent to

$$-9(6+8x)+12(1-4x)$$

50) Which expression is equivalent to

$$9(b-12)+4(-11-9b)$$

51) Which expression is equivalent to

$$-7(3r+10)-4(1+6r)$$

52) Which expression is equivalent to

$$7(9-8n)-7(n-5)$$

53) Which expression is equivalent to

$$-2(8+7m)+7(1+5m)$$

54) Which expression is equivalent to

$$8(6p+6)+11(p+5)$$

55) Which expression is equivalent to

$$-7(5a-8)-1(3a+4)$$

56) Which expression is equivalent to

$$12(-12-4k)-2(5-7k)$$

57) Which expression is equivalent to

$$-11(9-9x)-9(3x+10)$$

58) Which expression is equivalent to

$$-5(4n-5)-6(-9n-4)$$

59) Which expression is equivalent to

$$11(-2r+5)-4(-4r-1)$$

60) Which expression is equivalent to

$$-8(1+4r)+10(11-10r)$$

61)   Which expression is equivalent to

$$6 (9 p - 9) - 12 (4 + 11 p)$$

66)   Which expression is equivalent to

$$4 (x - 6) + 9 (x + 12)$$

62)   Which expression is equivalent to

$$-8 (6 x - 2) + 2 (- 2 x - 10)$$

67)   Which expression is equivalent to

$$4 (8 - 5 n) + 12 (n + 5)$$

63)   Which expression is equivalent to

$$10 (n - 3) + 8 (8 n + 5)$$

68)   Which expression is equivalent to

$$- 11 (- 8 n + 8) - 12 (10 - 4 n)$$

64)   Which expression is equivalent to

$$-9 (n - 10) - 7 (1 + 12 n)$$

69)   Which expression is equivalent to

$$5 (- 9 k + 5) + 9 (4 k - 11)$$

65)   Which expression is equivalent to

$$- 2 (p - 7) - 3 (2 p + 4)$$

70)   Which expression is equivalent to

$$- 11 (n + 9) + 8 (9 - 12 n)$$

71)   Which expression is equivalent to

$$4 (a + 9) + 10 (a - 9)$$

72)   Which expression is equivalent to

$$- 6 (9 n + 3) - 11 (6n - 4)$$

73)   Which expression is equivalent to

$$6 (1 + 8 x) + 10 (x - 6)$$

74)   Which expression is equivalent to

$$-2 (a + 4) - 10 (7 + 9 a)$$

75)   Which expression is equivalent to

$$-\frac{9}{4}\left(-\frac{1}{4}v + 1\right) + \frac{1}{5}$$

76)   Which expression is equivalent to

$$-\frac{6}{5}\left(r - \frac{1}{4}\right) + \frac{8}{3}$$

77)   Which expression is equivalent to

$$-\frac{8}{3} + \frac{5}{4}\left(-\frac{11}{4}b - \frac{7}{4}\right)$$

78)   Which expression is equivalent to

$$\frac{1}{3}m - \frac{1}{2}\left(-2m + \frac{7}{4}\right)$$

79)   Which expression is equivalent to

$$\frac{3}{5}\left(2p + \frac{1}{5}\right) + \frac{4}{5}p$$

80)   Which expression is equivalent to

$$-\frac{5}{2}n - 2\left(n - \frac{7}{2}\right)$$

81) Which expression is equivalent to

$$2 - \frac{5}{4}\left(\frac{3}{2}n + \frac{3}{2}\right)$$

82) Which expression is equivalent to

$$\frac{4}{3} - 2\left(-5v + \frac{2}{3}\right)$$

83) Which expression is equivalent to

$$\frac{2}{3}p + \frac{1}{2}\left(-\frac{3}{2}p + \frac{3}{2}\right)$$

84) Which expression is equivalent to

$$-\frac{13}{5}\left(x + \frac{5}{2}\right) - \frac{15}{4}x$$

85) Which expression is equivalent to

$$\frac{11}{4} + \frac{17}{4}\left(r - \frac{5}{2}\right)$$

86) Which expression is equivalent to

$$5\left(-n - \frac{7}{4}\right) + \frac{2}{3}$$

87) Which expression is equivalent to

$$-\frac{7}{4}b - \frac{5}{4}\left(\frac{3}{5}b + \frac{14}{5}\right)$$

88) Which expression is equivalent to

$$-\frac{7}{4} + \frac{5}{3}\left(x - \frac{7}{3}\right)$$

89) Which expression is equivalent to

$$-\frac{9}{4} + \frac{5}{4}\left(v - \frac{12}{5}\right)$$

90) Which expression is equivalent to

$$\frac{3}{2}\left(\frac{1}{2}n + \frac{5}{4}\right) - \frac{11}{5}n$$

91) Which expression is equivalent to

$$\frac{9}{4}\left(-\frac{8}{3}x + \frac{4}{5}\right) + 1$$

92) Which expression is equivalent to

$$\frac{1}{2}\left(x - \frac{13}{4}\right) + \frac{14}{5}$$

93) Which expression is equivalent to

$$-\frac{3}{2}\left(\frac{9}{4}n + \frac{3}{4}\right) + \frac{1}{4}n$$

94) Which expression is equivalent to

$$-x - 2\left(x - \frac{1}{2}\right)$$

95) Which expression is equivalent to

$$-\frac{7}{3}\left(b + \frac{5}{2}\right) - \frac{7}{4}$$

96) Which expression is equivalent to

$$-\frac{17}{5}\left(-\frac{1}{2}x + 2\right) - \frac{18}{5}x$$

97) Which expression is equivalent to

$$-\frac{18}{5}b + \frac{5}{3}\left(\frac{5}{4}b + \frac{3}{4}\right)$$

98) Which expression is equivalent to

$$\frac{4}{3} - \frac{3}{2}\left(\frac{1}{2}x - \frac{9}{5}\right)$$

99) Which expression is equivalent to

$$\frac{1}{2}\left(-\frac{3}{2}n - \frac{3}{2}\right) - \frac{5}{2}n$$

100) Which expression is equivalent to

$$-\frac{9}{4}\left(\frac{9}{5}p - 4\right) + 3$$

               www.math-knots.com | www.a4ace.com

101) Which expression is equivalent to

$$-\frac{11}{3} + \frac{2}{5}\left(2r + \frac{5}{4}\right)$$

102) Which expression is equivalent to

$$-\frac{4}{5}\left(\frac{2}{3}a + \frac{1}{4}\right) - \frac{4}{5}$$

1) Find the value of n :

$$-27 - n = -30$$

A) $\dfrac{10}{9}$ B) -57

C) -3 D) 3

2) Find the value of p :

$$17 = 8 - p$$

A) $\dfrac{17}{8}$ B) 9

C) -9 D) 25

3) Find the value of k :

$$39k = 429$$

A) -2 B) 22

C) 32 D) 11

4) Find the value of k :

$$-19 = k - 7$$

A) $-\dfrac{19}{7}$ B) 3

C) -12 D) -26

5) Find the value of x :

$$14x = 84$$

A) -15 B) 70

C) 98 D) 6

6) Find the value of x :

$$-12 = -2 - x$$

A) 10 B) 6

C) -10 D) -14

7) Find the value of r :

$$28 + r = 10$$

A) -18 B) $\dfrac{5}{14}$

C) -32 D) 38

8) Find the value of k :

$$-36 = \dfrac{k}{16}$$

A) -52 B) -20

C) -576 D) $-\dfrac{9}{4}$

9) Find the value of x :

$$-31 = x - 13$$

A) -22 B) -44

C) $-\dfrac{31}{13}$ D) -18

10) Find the value of a :

$$8 = \dfrac{a}{5}$$

A) 40 B) 13

C) 3 D) $\dfrac{8}{5}$

11) Find the value of n :

$$-37n = -1369$$

A) 37   B) −24

C) −3   D) 19

12) Find the value of m :

$$14 = \frac{m}{23}$$

A) 322   B) 37

C) $\frac{14}{23}$ D) −9

13) Find the value of n :

$$n + 23 = 22$$

A) −1   B) 45

C) 8   D) $\frac{22}{23}$

14) Find the value of x :

$$-4x = -104$$

A) 12   B) −36

C) −8   D) 26

15) Find the value of x :

$$-19 - x = 4$$

A) $-\frac{4}{19}$ B) −15

C) −23   D) 23

16) Find the value of k :

$$-42 = -15 + k$$

A) −27   B) $\frac{14}{5}$

C) 32   D) −57

17) Find the value of b :

$$-36 = b - 21$$

A) 38   B) −57

C) −15   D) $-\frac{12}{7}$

18) Find the value of a :

$$\frac{a}{20} = -24$$

A) −44   B) −4

C) $-\frac{6}{5}$ D) −480

19) Find the value of n :

$$1 = -31 + n$$

A) −31   B) $-\frac{1}{31}$

C) 32   D) −30

20) Find the value of n :

$$7 = -21 + n$$

A) −14   B) $-\frac{1}{3}$

C) 28   D) 29

21) Find the value of r :

$$52 = r + 14$$

   A) 24   B) 38

   C) 66   D) $\dfrac{26}{7}$

22) Find the value of x :

$$-1287 = 39x$$

   A) 23   B) 17

   C) 33   D) -33

23) Find the value of v :

$$-9 = v + 19$$

   A) -28 B) 10

   C) -1 D) $-\dfrac{9}{19}$

24) Find the value of n :

$$2 = n - 25$$

   A) 27   B) 50

   C) $\dfrac{2}{25}$ D) -23

25) Find the value of r :

$$-39 + r = -23$$

   A) -62 B) 24

   C) 16   D) $\dfrac{23}{39}$

26) Find the value of x :

$$7 = 16 + x$$

   A) $\dfrac{7}{16}$ B) 29

   C) -9 D) 23

27) Find the value of x :

$$8 = \dfrac{x}{34}$$

   A) 272   B) $\dfrac{4}{17}$

   C) -26 D) 42

28) Find the value of x :

$$-33 = 6 - x$$

   A) $-\dfrac{11}{2}$ B) 39

   C) -27 D) -39

29) Find the value of x :

$$x + 12 = 2$$

   A) 14   B) 24

   C) $\dfrac{1}{6}$ D) -10

30) Find the value of x :

$$8 = 7 - x$$

   A) -1 B) 15

   C) 1   D) $\dfrac{8}{7}$

          www.math-knots.com | www.a4ace.com

31) Find the value of n :

$$-3 = n + 35$$

A) −13   B) 32

C) $-\dfrac{3}{35}$ D) −38

32) Find the value of r :

$$r - 27 = -43$$

A) $-\dfrac{43}{27}$ B) −16

C) −70   D) 23

33) Find the value of r :

$$-41 = r - 19$$

A) −60   B) −22

C) $-\dfrac{41}{19}$ D) 8

34) Find the value of x :

$$4x = 60$$

A) 15   B) 18

C) 56   D) 64

35) Find the value of x :

$$\dfrac{x}{15} = -37$$

A) $-\dfrac{37}{15}$ B) −555

C) −52   D) −22

36) Find the value of v :

$$-16 = v + (-37)$$

A) $\dfrac{16}{37}$ B) −53

C) 21   D) −39

37) Find the value of m :

$$-26 = m - 22$$

A) −4   B) 38

C) −48   D) $-\dfrac{13}{11}$

38) Find the value of r :

$$r - (-22) = -10$$

A) 12   B) 25

C) −32   D) $\dfrac{5}{11}$

39) Find the value of x :

$$-59 = x + (-24)$$

A) $\dfrac{59}{24}$ B) −83

C) −35   D) 34

40) Find the value of x :

$$\dfrac{x}{17} = \dfrac{33}{17}$$

A) 4   B) −8

C) −10   D) 33

       www.math-knots.com | www.a4ace.com

41) Find the value of x :

$$805 = 23\,x$$

A) −25 B) 25

C) 35 D) 11

42) Find the value of x :

$$\frac{x}{37} = -23$$

A) −60 B) $-\frac{23}{37}$

C) 14 D) −851

43) Find the value of x :

$$\frac{x}{30} = 33$$

A) $\frac{11}{10}$ B) 3

C) 63 D) 990

44) Find the value of b :

$$20 - b = 42$$

A) 22 B) $\frac{21}{10}$

C) −22 D) 62

45) Find the value of n :

$$-18\,n = -432$$

A) −32 B) 24

C) 32 D) −37

46) Find the value of m :

$$13\,m = -234$$

A) 6 B) - 34

C) −18 D) −20

47) Find the value of x :

$$29.5\,x = 501.5$$

A) 11.7 B) 27.22

C) 1.8 D) 17

48) Find the value of v :

$$24 + v = 44.5$$

A) 20.5 B) 14.028

C) −29.2 D) 11.6

49) Find the value of x :

$$\frac{x}{7.2} = -28.7$$

A) −0.52 B) −206.64

C) 28.9 D) −36

50) Find the value of x :

$$195.65 = -6.5\,x$$

A) −30.1 B) −36.7

C) 14.9 D) 6.5

www.math-knots.com | www.a4ace.com

51) Find the value of n :

$-23.6\,n = -396.48$

A) 24.3   B) 7.2

C) 16.8   D) 26.4

52) Find the value of x :

$-9\,x = -46.8$

A) −17.5   B) −18.4

C) 5.2   D) 37.335

53) Find the value of n :

$\dfrac{n}{36.3} = -39.1$

A) −2.8   B) 13.3

C) −1419.33 D) −14.1

54) Find the value of x :

$-15.4 = x + (-16.2)$

A) 12   B) 19.1

C) 0.8   D) 31.9

55) Find the value of n :

$-40.62 = n - 8.8$

A) −14.3   B) 32.8

C) −31.82   D) −22

56) Find the value of x :

$x - 19.3 = 5.637$

A) −24.2   B) 24.937

C) −38.3   D) 22.68

57) Find the value of r :

$r - (-6.93) = 37.33$

A) 30.4   B) −16.4

C) 6.1   D) −38.9

58) Find the value of a :

$19.4\,a = 428.74$

A) 7.6   B) 22.1

C) 37.005   D) 17.2

59) Find the value of r :

$-184.15 = 14.5\,r$

A) 39.6   B) −8.394

C) 24   D) −12.7

60) Find the value of x :

$x + (-34) = -12.3$

A) −18.6   B) 0.5

C) 21.7   D) 19.6

61) Find the value of n :

$$-46.5 = -23.7 + n$$

A) 14.2   B) −22.8

C) −11.9   D) −13.5

62) Find the value of k :

$$-27.6 = \frac{k}{0.8}$$

A) −20.9   B) 18.5

C) 15.5   D) −22.08

63) Find the value of p :

$$7(1 + 6p) = -119$$

64) Find the value of m :

$$-7(6m - 7) + 5m = 345$$

65) Find the value of v :

$$-114 = -6(5 - 2v)$$

66) Find the value of n :

$$149 = 2(8n + 7) + 7$$

67) Find the value of x :

$$-264 = -6(8x - 4)$$

68) Find the value of k :

$$-83 = 5(-5 + 6k) - k$$

69) Find the value of a :

$$-7 + 5(-6a - 2) = -107$$

70) Find the value of r :

$$-2(-5 - 4r) + 2r = 90$$

71)   Find the value of x :

$$105 = 1 + 8 (3 + 2 x)$$

72)   Find the value of r :

$$- 88 = - 4 (- 3 r + 1)$$

73) Find the value of x :

$$- 8 (8 x + 7) + 7 x = 286$$

74) Find the value of n :

$$- 3 (8 + 6 n) = - 96$$

75) Find the value of n :

$$- 6 (- 6 + 4 n) = - 132$$

76) Find the value of m :

$$- 3 (1 - 3 m) + 5 m = - 115$$

77) Find the value of n :

$$134 = - 2 (- 3 - 8 n)$$

78)   Find the value of v :

$$5 (- 2 v + 5) = 95$$

79)   Find the value of x :

$$- 88 = 6 (2 x - 4) + 8$$

80)   Find the value of n :

$$- 272 = 8 (1 - 7 n)$$

81) Find the value of x :

$$250 = 5 (2 - 8 x)$$

82) Find the value of p :

$$90 = - 3 (2 - 5 p) + 6$$

83) Find the value of n :

$$- 280 = 7 (6 n + 2)$$

84) Find the value of n :

$$- 108 = - 3 (7 n + 8)$$

85) Find the value of a :

$$100 = 2 (- 6 - 7 a)$$

86) Find the value of x :

$$- 90 = - 2 (7 x + 5) + 4 x$$

87) Find the value of b :

$$288 = 8 (1 - 5 b)$$

88) Find the value of p :

$$6 (5 - 7 p) = - 264$$

89) Find the value of v :

$$- 4 (6 v + 5) = 148$$

1) Find the value of n :

$$90 - 2n = -2(1 + 6n) - 13n$$

A) −4   B) −26

C) No solution.   D) −34

2) Find the value of m :

$$10(13 - 7m) = -2 - 4m$$

A) −39   B) 26

C) 2   D) −37

3) Find the value of x :

$$-18 + 14x = 18(x - 5)$$

A) 18   B) No solution.

C) −7   D) 20

4) Find the value of r :

$$-76 - 19b = 6(-15b + 11)$$

A) −36   B) −19

C) 2   D) −3

5) Find the value of x :

$$95 + 18x = -7(1 - 5x)$$

A) 6   B) 36

C) 34   D) −13

6) Find the value of p :

$$15(p + 20) = 2p + 66$$

A) 1   B) −18

C) No solution.   D) −5

7) Find the value of x :

$$13(11x - 2) = -26 + 13x$$

A) −38   B) 0

C) No solution.   D) 5

8) Find the value of n :

$$5(n - 7) = -23 - n$$

A) −17   B) −34

C) −13   D) 2

9) Find the value of r :

$$-7r - 25 = -4 + 14(3r - 12)$$

A) −17   B) 3

C) 30   D) 36

10) Find the value of n:

$$-14(n + 16) = 13n - 89$$

A) −10   B) −19

C) −14   D) −5

                   www.math-knots.com | www.a4ace.com

11) Find the value of r :

$$84 - 3r = 16(r - 9)$$

A) 8
B) 12
C) All real numbers.
D) 15

12) Find the value of v :

$$-6v + 29 = -5 - 14(v + 1)$$

A) −40
B) −36
C) −6
D) All real numbers.

13) Find the value of b :

$$5(13b + 15) = 75 + 4b$$

A) 11
B) All real numbers.
C) −14
D) 0

14) Find the value of x :

$$2(1 + x) = 18x - 78$$

A) 8   B) All real numbers.

C) 5   D) 29

15) Find the value of a :

$$-30 + 3a = -18(4a - 15)$$

A) 12   B) 4

C) −20   D) No solution.

16) Find the value of v :

$$-(2v - 9) = 57 + 4v$$

A) −8   B) 25

C) 15   D) −27

17) Find the value of n :

$$6(2n + 15) = 6n$$

A) 22   B) −15

C) −8   D) 35

18) Find the value of m :

$$-17(1 - 2m) = -92 + 19m$$

A) −25   B) −5

C) −34   D) 24

19) Find the value of k :

$$-65 - 17k = 17(k - 10) - 13k$$

A) 36   B) −19

C) −5   D) −7

20) Find the value of x:

$$-4(1 - 12x) = 5x - 90$$

A) No solution.   B) −38

C) −2   D) −35

Algebra 1

21) Find the value of x :

$$39 + 7x = 10x - 20(x - 13)$$

A) 13   B) 7

C) 24   D) -39

22) Find the value of n :

$$-76 - 9n = -2(n - 18) - 14n$$

A) 23   B) -21

C) 16   D) -3

23) Find the value of b :

$$-5v - 3(13 + 7v) = 89 - 18v$$

A) -1   B) -16

C) -20   D) 23

24) Find the value of x :

$$3(7x - 4) = -68 + 7x$$

A) All real numbers.
B) No solution.
C) 19
D) -4

25) Find the value of x :

$$-19(17 + x) = -4x - 53$$

A) -18   B) 6

C) No solution.   D) 36

26) Find the value of m :

$$-9(m + 3) - 2 = -5 - 11m$$

A) 12   B) 4

C) -30   D) -27

27) Find the value of b :

$$-52 - 7b = -9(b + 10)$$

A) -18   B) -19

C) 12   D) -1

28) Find the value of n :

$$-10(n - 16) = -61 + 3n$$

A) No solution.
B) All real numbers.
C) -36
D) 17

29) Find the value of r :

$$-12(10r - 17) = -52 + 8r$$

A) 2
B) No solution.
C) All real numbers.
D) -14

30) Find the value of x:

$$-4(-6x + 12) + 2(1 + 4x) = 114$$

A) 25
B) All real numbers.
C) 18
D) 5

www.math-knots.com | www.a4ace.com

31) Find the value of n :

$$23 = -13(2 + 3n) + 2(1 - 4n)$$

A) −21   B) 2

C) 21   D) −1

32) Find the value of x :

$$9(x - 10) - 14(-6x + 6) = 12$$

A) 25   B) 2

C) −11   D) 26

33) Find the value of p :

$$-32 = -7(4 + 6p) + 11(8p + 8)$$

A) −2   B) 16

C) −14   D) 24

34) Find the value of r :

$$-23 = -11(-7r + 12) + 8(11r - 7)$$

A) 18   B) 15

C) 1   D) −25

35) Find the value of x :

$$32 = 14(2 - 9x) + 4(1 + 4x)$$

A) 25   B) 0

C) 3   D) 2

36) Find the value of r :

$$-1 = 2(r - 12) - 7(-11 + 8r)$$

A) All real numbers.
B) 4
C) 1
D) No solution.

37) Find the value of b :

$$-115 = -14(b + 5) - 9(5 + 13b)$$

A) −5   B) 0

C) 17   D) −19

38) Find the value of n :

$$34 = -8(10n - 3) + 10(9n + 6)$$

A) −5   B) −20

C) No solution.   D) 4

39) Find the value of b :

$$8(b - 1) + 7(-4b + 8) = 28$$

A) −4   B) −21

C) 1   D) 17

40) Find the value of b:

$$-9(-2b - 6) - 14(-1 - 3b) = -112$$

A) No solution.
B) −3
C) −21
D) All real numbers.

www.math-knots.com | www.a4ace.com

 **Algebra 1**

41) Find the value of a :

$-140 = 12 (14 + 6 a) - 4 (a - 8)$

A) −5   B) 26

C) −3   D) 27

42) Find the value of b :

$4 (- 14 + 11 b) - 3 (1 + 2 b) = 55$

A) 3   B) −10

C) −5   D) 28

43) Find the value of x :

$-53 = 9 (7 x - 4) - 8 (- 4 x + 14)$

A) 1   B) −2

C) 8   D) −21

44) Find the value of k :

$-46 = 8 (14 k - 4) + 14 (- 1 - 8 k)$

A) All real numbers.
B) No solution.
C) −3
D) 27

45) Find the value of p :

$-3 (4 p + 3) + 5 (p - 5) = - 13$

A) −3   B) −6

C) No solution.   D) 18

46) Find the value of x :

$-8 (3 x + 13) + 6 (10 x - 4) = - 92$

A) 18   B) No solution.

C) 1   D) 20

47) Find the value of m :

$-3 (m - 4) + 7 (5 m + 6) = 86$

A) 18   B) 10

C) 0   D) 1

48) Find the value of m :

$2 (4 m + 13) - 3 (5 m - 10) = 140$

A) −10   B) −12

C) No solution.   D) −15

49) Find the value of x :

$-87 = 9 (x - 6) + 10 (- 4 x + 6)$

A) No solution.   B) −12

C) −17   D) 3

50) Find the value of p:

$4 (p + 1) + 14 (1 - 13 p) = 18$

A) 28   B) 0

C) −11   D) 12

51) Find the value of k :

$-53 = -3(9k + 14) + 10(13 - 2k)$

A) −12   B) 15

C) −23   D) 3

52) Find the value of n :

$-(-12n - 14) - 11(8 + n) = -66$

A) − 24   B) −15

C) − 20   D) 8

53) Find the value of m :

$2(5m - 6) - 12(m - 8) = 68$
A) 10
B) 8
C) 20
D) All real numbers.

54) Find the value of b :

$-7(7b - 1) + 9(6b + 11) = 86$

A) −4   B) −9

C) −28   D) −8

55) Find the value of m :

$5(9 - 7m) + 6(m - 12) = 31$

A) 11   B) − 2

C) − 16   D) − 9

56) Find the value of x :

$13(-2x - 6) - 9(-11 + 10x) = -95$

A) 13   B) 1

C) − 10   D) 5

57) Find the value of x :

$-12(x + 7) + 6(13x + 8) = -36$

A) 0   B) 25

C) − 2   D) − 14

58) Find the value of v :

$-3 = 9(1 - 4v) - 12(1 - 3v)$

A) No solution.
B) All real numbers.
C) 10
D) 25

59) Find the value of n :

$2 = 14(1 + 6n) + 12(12 + 6n)$

A) 21
B) −1
C) −24
D) All real numbers.

60) Find the value of x:

$4x - (12x + 12) = -3(x + 12) - 8x$

A) 19   B) − 8

C) − 23   D) 0

www.math-knots.com | www.a4ace.com

61) Find the value of a :

$5 (4 + 3 a) - 4 = 8 (1 + 2 a)$

A) 24
B) −21
C) 8
D) All real numbers.

62) Find the value of b :

$- 9 (9 b - 6) = 11 (- 6 b - 6)$

A) No solution. B) −15
C) 2 D) 8

63) Find the value of k :

$- 4 (1 - 5 k) = 4 (4 k + 7)$

A) 8 B) 1

C) 17 D) −11

64) Find the value of k :

$12 (6 k - 11) = - 4 (11 k + 4)$

A) −3
B) All real numbers.
C) No solution.
D) 1

65) Find the value of k :

$- 5 (7 - 10 k) + 3 = 7 (1 + 9 k)$

A) − 20 B) − 4

C) 15 D) − 3

66) Find the value of v :
$v + 11 + 9 - 7 v =$
$4 (8 v - 10) + 6 (10 - 12 v)$

A) 2 B) −3

C) 0 D) No solution.

67) Find the value of b :

$8 (9 - b) + 4 = 8 b + 4 (b + 4)$

A) 0 B) − 23

C) 18 D) 3

68) Find the value of x :

$- 12 (x - 6) - 4 (- 4 - 10 x)$
$= - 12 x - 4 x$

A) −22 B) −8

C) 15 D) −2

69) Find the value of r :
$3 (5 - r) = - 8 r + 2 (2 r + 6)$

A) 15
B) − 22
C) − 3
D) No solution.

70) Find the value of p:
$11 - 2 (3 p + 12)$
$= 4 (p - 3) - 9 p$

A) 7 B) −1

C) −19 D) No solution.

71) Find the value of x :

$$-11(-2+8x)=6(-12x+9)$$

A) −23
B) −2
C) All real numbers.
D) 0

72) Find the value of p :

$$-8(p-7)=-4(p+7)$$

A) −21   B) 21

C) 8   D) −9

73) Find the value of n :

$$10(n-6)=10(n+10)$$

A) 7   B) No solution.

C) 8   D) −17

74) Find the value of x :

$$3(x+11)=-7(2x-8)+11$$

A) −19   B) 3

C) 2   D) −18

75) Find the value of a :

$$-8a-11(1-10a)=7(1+12a)$$

A) 1   B) − 4

C) − 13   D) − 21

76) Find the value of v :

$$9(v+3)=-9(v-8)+9v$$

A) No solution.   B) −3

C) 5   D) −14

77) Find the value of p :

$$6(8-11p)-9(-10p+6)$$
$$=12p+11+5p+11$$

A) − 18   B) 7

C) 4   D) 18

78) Find the value of n :

$$2(11n+9)-4n=2(11n-3)$$

A) 18   B) 15

C) 6   D) 0

79) Find the value of p :

$$-3(p-12)-6p=-3(p-10)$$

A) 7   B) 18

C) 1   D) −3

80) Find the value of n:

$$-3(n+5)=-7(n-11)$$

A) −7
B) 23
C) All real numbers.
D) −13

81) Find the value of n :

$$-15 (3 + 5 n) + 11 n = -10 n - 5 (n - 1)$$

A) 2   B) 7

C) 17   D) -1

82) Find the value of r :

$$-5 (1 - 5 r) = 5 (3 r - 7)$$

A) −3   B) 24

C) 19   D) − 16

83) Find the value of x :

$$-9 + 3 (x + 9) = 3 x - 3 (4 - 4 x)$$

A) −18   B) −4

C) 12   D) 5

84) Find the value of n :

$$11 (1 - 3 n) = 5 (- 8 n - 3) + 7 n$$

A) 13   B) 19

C) No solution.   D) −2

85) Find the value of m :

$$-3 (- 7 m - 6) = 8 (2 m - 2) - 12 m$$

A) − 24   B) 21

C) 15   D) − 2

www.math-knots.com | www.a4ace.com

 Algebra 1

1) Find the value of 'v' that satisfies the below equation.

$$15.69 = -6.3 v + 6.9 - 5.7$$

A) −4.9   B) 0.1

C) 8.6   D) −2.3

2) Find the value of 'n' that satisfies the below equation.

$$-2.3 n + 5.1 - 5.8 = 7.81$$

A) −12.96   B) 3

C) −3.8   D) −3.7

3) Find the value of 'x' that satisfies the below equation.

$$-1.95 = 1 - 0.4 x - 2.99$$

A) 8.7   B) 5.2

C) −0.1   D) −4.1

4) Find the value of 'm' that satisfies the below equation.

$$-1.8 m + 3.5 m = 6.8$$

A) 4   B) 9.9

C) 10.7   D) 9.35

5) Find the value of 'x' that satisfies the below equation.

$$x + 2.9 + 3.72 = 8.06$$

A) 14.7   B) 1.44

C) −7.06   D) −0.1

6) Find the value of 'x' that satisfies the below equation.

$$-20.66 = 1 + 7.8 x + 3.3$$

A) 3.42   B) 13.4

C) 15.7   D) −3.2

7) Find the value of 'p' that satisfies the below equation.

$$-2.26 p + 2.3 + 5.6 p = -7.72$$

A) −3   B) 9.3

C) 13   D) −0.1

8) Find the value of 'a' that satisfies the below equation.

$$1 - 6.7 a - 2.8 a = 3.66$$

A) −4.1
B) All real numbers.
C) No solution.
D) −0.28

9) Find the value of 'x' that satisfies the below equation.

$$6.8 x - 5.91 - 3.1 = -15.13$$

A) 3.22   B) −0.24

C) −4.3   D) −0.9

10) Find the value of 'n' that satisfies the below equation.

$$-5.1 n - 7.5 n = -3.78$$

A) 9.9   B) 2.2

C) 0.5   D) 0.3

11) Find the value of 'x' that satisfies the below equation.

$$-17.25 = 4.9\,x - 4.3 + 4.2$$

A) 0.3   B) −5.5

C) 13.6   D) −3.5

12) Find the value of 'x' that satisfies the below equation.

$$1 - 3.4\,x - 4.8\,x = 11.66$$

A) −10.3
B) 9.9
C) −1.3
D) All real numbers.

13) Find the value of 'x' that satisfies the below equation.

$$-6.7\,x - 2.1\,x = -16.72$$

A) 9.6   B) 1.9

C) −8.6   D) 4.6

14) Find the value of 'r' that satisfies the below equation.

$$-6.6\,r + 3.6\,r = 20.13$$

A) 0.8   B) 9.5

C) −6.71   D) 6.9

15) Find the value of 'p' that satisfies the below equation.

$$0.2\,p + 0.1\,p = 0.96$$

A) −2.3   B) −14

C) 3.2   D) No solution.

16) Find the value of 'a' that satisfies the below equation.

$$-6.18 = a + 7.4 - 5.8$$

A) −7.78   B) −7.7

C) No solution. D) 6.5

17) Find the value of 'b' that satisfies the below equation.

$$-1 = 0.4\,b - 0.4\,b$$

A) 12.7
B) 0.7
C) No solution.
D) All real numbers.

18) Find the value of 'k' that satisfies the below equation.

$$-15.07 = -7.1\,k - 6.6\,k$$

A) −1.7   B) −7.88

C) 1.1   D) −5.7

19) Find the value of 'p' that satisfies the below equation.

$$p + 4.5 - 6.8 = -5.7$$

A) −3.4   B) No solution.

C) 4.047   D) 13

20) Find the value of 'n' that satisfies the below equation.

$$-20.21 = 0.1 - 4.55\,n - 0.29$$

A) −12.3   B) 7.4

C) −1.6   D) 4.4

 **Algebra 1**

21) Find the value of 'n' that satisfies the below equation.

$$-3.8\,n + 2.5\,n = 7.28$$

A) −11.3   B) 14.27

C) −5.6   D) No solution.

22) Find the value of 'x' that satisfies the below equation.

$$7x + 6.6 - 0.6x = -18.36$$

A) 15.4   B) −3.9

C) No solution.   D) −1.7

23) Find the value of 'x' that satisfies the below equation.

$$-1.6x + 4x = 18$$

A) 7.5   B) −5.7

C) −12.3   D) −5.5

24) Find the value of 'x' that satisfies the below equation.

$$1.2 - 3.9x + 1.1 = 8.15$$

A) −6.5   B) 6.9

C) −1.5   D) −12.3

25) Find the value of 'r' that satisfies the below equation.

$$0 = 4.8\,r - 4.8\,r$$

A) −2.8
B) All real numbers.
C) 6.6
D) −7.4

26) Find the value of 'm' that satisfies the below equation.

$$3.9 = m - 4.9 + 1.1$$

A) −2.3   B) 14

C) 11.1   D) 7.7

27) Find the value of 'k' that satisfies the below equation.

$$-\frac{49}{48} = \frac{19}{6}k - \frac{9}{8}k$$

A) $-\frac{11}{9}$ B) $2\frac{2}{15}$

C) $\frac{4}{3}$ D) $-\frac{1}{2}$

28) Find the value of 'b' that satisfies the below equation.

$$0 = -b + b$$

A) $\frac{1}{8}$ B) $-\frac{1}{7}$

C) All real numbers. D) $4\frac{3}{4}$

29) Find the value of 'x' that satisfies the below equation.

$$-\frac{5}{3}x + \frac{1}{2} + 4\frac{3}{4} = -\frac{193}{28}$$

A) $\frac{51}{7}$ B) $-\frac{5}{4}$

C) No solution.   D) $\frac{3}{2}$

30) Find the value of 'x' that satisfies the below equation.

$$\frac{2}{3}x + \frac{3}{2} - \frac{3}{2}x = \frac{4}{21}$$

A) $3\frac{1}{2}$

B) All real numbers.

C) $\frac{16}{15}$

D) $\frac{11}{7}$

         www.math-knots.com | www.a4ace.com

31) Find the value of 'x' that satisfies the below equation.

$$0 = \frac{1}{2}x - \frac{1}{2}x$$

A) $1\frac{5}{16}$

B) $-\frac{21}{13}$

C) $-1\frac{5}{8}$

D) All real numbers.

32) Find the value of 'x' that satisfies the below equation.

$$\frac{25}{16} = x + \frac{1}{2} + \frac{9}{8}x$$

A) All real numbers.

B) $2\frac{6}{7}$

C) $\frac{7}{15}$

D) $\frac{1}{2}$

33) Find the value of 'x' that satisfies the below equation.

$$-\frac{7}{10} = \frac{14}{5}x + 1 + 3\frac{1}{5}$$

A) $-2\frac{1}{6}$ B) $-\frac{7}{4}$

C) 4   D) -2

34) Find the value of 'v' that satisfies the below equation.

$$\frac{10831}{840} = \frac{19}{5}v + \frac{9}{7} + \frac{17}{6}v$$

A) $\frac{15}{8}$ B) $4\frac{11}{15}$

C) $\frac{7}{4}$ D) $5\frac{3}{8}$

35) Find the value of 'x' that satisfies the below equation.

$$-\frac{175}{24} = -2x - \frac{23}{6}x$$

A) $-\frac{8}{15}$ B) $\frac{5}{4}$

C) $3\frac{1}{2}$ D) $2\frac{4}{15}$

36) Find the value of 'b' that satisfies the below equation.

$$-\frac{671}{126} = -\frac{11}{7}b - \frac{4}{3}b$$

A) $7\frac{1}{8}$ B) $1\frac{1}{11}$

C) $\frac{1}{3}$ D) $\frac{11}{6}$

37) Find the value of 'n' that satisfies the below equation.

$$\frac{29}{42} = \frac{1}{7}n - \frac{1}{6} + \frac{1}{7}$$

A) $-\frac{5}{4}$

B) All real numbers.

C) 5

D) $4\frac{7}{12}$

38) Find the value of 'v' that satisfies the below equation.

$$\frac{187}{336} = \frac{6}{7}v + 2 - \frac{19}{6}v$$

A) $3\frac{3}{7}$ B) 0

C) $-1\frac{1}{7}$ D) $\frac{5}{8}$

39) Find the value of 'n' that satisfies the below equation.

$$\frac{3}{8}n + \frac{9}{4} - 3\frac{1}{2} = -\frac{3}{4}$$

A) 2   B) $\frac{4}{3}$

C) $-\frac{5}{14}$ D) $-5\frac{8}{11}$

40) Find the value of 'x' that satisfies the below equation.

$$\frac{10}{3}x + \frac{37}{8} + \frac{1}{4}x = 10$$

A) $\frac{4}{3}$ B) $\frac{1}{2}$

C) $-2\frac{9}{10}$ D) $\frac{3}{2}$

41) Find the value of 'r' that satisfies the below equation.

$$-\frac{117}{16} = 2r - \frac{29}{8}r$$

A) $\frac{1}{4}$ B) $\frac{3}{4}$

C) No solution.   D) $\frac{9}{2}$

42) Find the value of 'r' that satisfies the below equation.

$$-\frac{6}{5}r - \frac{3}{2}r = -\frac{9}{20}$$

A) $\frac{5}{4}$ B) $\frac{17}{9}$

C) -10 D) $\frac{1}{6}$

www.math-knots.com | www.a4ace.com

43) Find the value of 'a' that satisfies the below equation.

$$-\frac{5}{3} - 4 = -\frac{5}{4}a - \frac{5}{3} + \frac{5}{4}a$$

A) $-\frac{2}{3}$ B) No solution.

C) $-\frac{1}{2}$ D) $3\frac{1}{3}$

44) Find the value of 'v' that satisfies the below equation.

$$-\frac{8}{7}v - \frac{5}{4}v = \frac{67}{20}$$

A) No solution.   B) $-2\frac{11}{15}$

C) $3\frac{1}{2}$ D) $-\frac{7}{5}$

45) Find the value of 'b' that satisfies the below equation.

$$-\frac{3}{4}b - 7b = -\frac{31}{3}$$

A) $-\frac{17}{12}$ B) $\frac{12}{13}$

C) $\frac{7}{10}$ D) $\frac{4}{3}$

46) Find the value of 'x' that satisfies the below equation.

$$\frac{2}{7}x - \frac{3}{2}x = \frac{17}{7}$$

A) All real numbers.

B) $\frac{11}{12}$

C) $-\frac{8}{5}$

D) -2

47) Find the value of 'v' that satisfies the below equation.

$$\frac{13}{3}v + 2 - \frac{3}{4}v = -\frac{23}{10}$$

A) All real numbers.

B) $1\frac{11}{13}$

C) -2

D) $-\frac{6}{5}$

48) Find the value of 'b' that satisfies the below equation.

$$\frac{107}{28} = -\frac{9}{8}b + \frac{11}{7} - \frac{3}{4}$$

A) $\frac{2}{3}$ B) -8

C) $-\frac{8}{3}$ D) $\frac{11}{13}$

49) Find the value of 'x' that satisfies the below equation.

$$\frac{5}{4} = x + \frac{5}{4} - x$$

A) $\frac{15}{16}$

B) -3

C) All real numbers.

D) $\frac{13}{8}$

50) Find the value of 'n' that satisfies the below equation.

$$-n + \frac{12}{7}n = \frac{20}{7}$$

A) $\frac{6}{7}$

B) $-2\frac{1}{12}$

C) $-2\frac{3}{13}$

D) 4

51) Find the value of 'n' that satisfies the below equation.

$$-\frac{7}{4}n + \frac{9}{5} + 4\frac{5}{8} = \frac{759}{80}$$

A) $8\frac{1}{2}$

B) $-\frac{1}{4}$

C) $2\frac{5}{12}$

D) $-\frac{7}{4}$

52) Find the value of 'b' that satisfies the below equation.

$$-7b - \frac{3}{7}b = 0$$

A) $-2\frac{4}{9}$

B) $-\frac{1}{2}$

C) 0

D) $\frac{22}{13}$

53) Find the value of 'x' that satisfies the below equation.

$$-\frac{4}{5}x + \frac{7}{4} - \frac{5}{4} = \frac{59}{30}$$

A) $-3\frac{1}{3}$

B) $-\frac{11}{6}$

C) $-\frac{11}{10}$

D) $-\frac{1}{3}$

54) Find the value of 'v' that satisfies the below equation.

$$\frac{341}{150} = \frac{9}{5}v - \frac{17}{6}v$$

A) $\frac{1}{12}$

B) $-\frac{1}{11}$

C) $-\frac{11}{5}$

D) $\frac{13}{8}$

55) Find the value of 'x' that satisfies the below equation.

$$7x + 3(5 + 7x) = 211$$

A) 0   B) −1

C) 9   D) 7

56) Find the value of 'm' that satisfies the below equation.

$$-170 = -8 + 6(1 + 7m)$$

A) −4   B) 9

C) −2   D) 0

57) Find the value of 'x' that satisfies the below equation.

$$6x - 4(1 + 5x) = 94$$

A) 11   B) 12

C) −7   D) 15

58) Find the value of 'm' that satisfies the below equation.

$$100 = 2(5m + 6) + m$$

A) −9   B) 9

C) −4   D) 8

59) Find the value of 'n' that satisfies the below equation.

$$-n + 5(3 - 7n) = 159$$

A) 5   B) 13

C) −4   D) No solution.

60) Find the value of 'k' that satisfies the below equation.

$$-99 = 7k - 8(-6 + 7k)$$

A) −6   B) 3

C) −14   D) 1

61) Find the value of 'b' that satisfies the below equation.

$$-6b - 4(5b + 6) = -206$$

A) −2   B) −15

C) 12   D) 7

62) Find the value of 'n' that satisfies the below equation.

$$-5 + 6(7n + 5) = -311$$

A) No solution.
B) All real numbers.
C) 15
D) −8

63) Find the value of 'a' that satisfies the below equation.

$$-6(-5 - 5a) = -180$$

A) −7   B) 1

C) 0   D) 14

64) Find the value of 'x' that satisfies the below equation.

$$-8(3 + 5x) = -184$$

A) −9   B) 9

C) 4   D) 16

65) Find the value of 'r' that satisfies the below equation.

$$-125 = 5(4r + 7)$$

A) −9   B) 7

C) −1   D) −8

66) Find the value of 'n' that satisfies the below equation.

$$-272 = 8(1 - 7n)$$

A) 5   B) 0

C) 9   D) No solution.

67) Find the value of 'm' that satisfies the below equation.

$$-8(-m - 3) = 88$$

A) −6
B) All real numbers.
C) No solution.
D) 8

68) Find the value of 'r' that satisfies the below equation.

$$6(3r + 8) = 174$$

A) 16   B) −14

C) −5   D) 7

69) Find the value of 'n' that satisfies the below equation.

$$6(6 - 4n) = 180$$

A) No solution.   B) 5

C) −6   D) −3

70) Find the value of 'v' that satisfies the below equation.

$$147 = 2 + 5(1 + 7v)$$

A) 1   B) 4

C) No solution.   D) −3

71) Find the value of 'x' that satisfies the below equation.

$$245 = 5(1 - 8x)$$

A) −12   B) −6

C) −7   D) 0

72) Find the value of 'p' that satisfies the below equation.

$$-112 = 7(3p + 5)$$

A) −5   B) 12

C) 3   D) −7

73) Find the value of 'r' that satisfies the below equation.

$$96 = 3(8 - 4r) + 3r$$

A) −16   B) −8

C) −6   D) 2

74) Find the value of 'p' that satisfies the below equation.

$$-88 = -2(5p + 4)$$

A) 3
B) 8
C) No solution.
D) All real numbers.

75) Find the value of 'a' that satisfies the below equation.

$$-96 = -8(2a - 4)$$

A) −15   B) −4

C) 8   D) −16

76) Find the value of 'r' that satisfies the below equation.

$$81 = 3(3 - 3r)$$

A) No solution.
B) All real numbers.
C) −8
D) 1

77) Find the value of 'x' that satisfies the below equation.

$$6x - 7(-4x - 6) = 110$$

A) 2   B) −6

C) −7   D) −2

78) Find the value of 'p' that satisfies the below equation.

$$7(1 - 5p) + 3p = -121$$

A) All real numbers.
B) 11
C) 1
D) 4

79) Find the value of 'x' that satisfies the below equation.

$$-127 = -7 + 8(-7 - 4x)$$

A) 2   B) −14

C) −9   D) 9

80) Find the value of 'x' that satisfies the below equation.

$$-4(5x + 2) - 4x = -200$$

A) −5   B) −15

C) No solution.   D) 8

81) Find the value of 'x' that satisfies the below equation.

$$-148 = x - 3(6x + 4)$$

A) 13   B) −3

C) −12   D) 8

82) Find the value of 'k' that satisfies the below equation.

$$-5(8 - 2k) = -90$$

A) −9   B) No solution.

C) 15   D) −5

83) Find the value of 'b' that satisfies the below equation.

$$6(4b - 7) = 102$$

A) −2   B) −7

C) 6   D) −3

84) Find the value of 'r' that satisfies the below equation.

$$-7(r + 5) = -84$$

A) No solution.   B) 11

C) −11   D) 7

85) Find the value of 'x' that satisfies the below equation.

$7 (x + 4) + 8 (1 - 7 x) = - 62$

A) All real numbers.
B) −16
C) 1
D) 2

86) Find the value of 'n' that satisfies the below equation.

$7 (1 - 5 n) - 3 (n - 1) = - 66$

A) −6   B) 1

C) 8   D) 2

87) Find the value of 'k' that satisfies the below equation.

$- 4 (4 + 8 k) + 8 (2 k + 4) = - 32$

A) 13
B) All real numbers.
C) 3
D) 7

88) Find the value of 'v' that satisfies the below equation.

$- 8 (1 - 7 v) - (v + 1) = - 9$

A) 12   B) 0

C) 3   D) −7

89) Find the value of 'x' that satisfies the below equation.

$- 8 (- 7 x + 1) + 7 (x - 5) = - 43$

A) −10   B) 3

C) 14   D) 0

90) Find the value of 'n' that satisfies the below equation.

$4 (1 - 5 n) - 2 (6 n + 1) = 66$

A) −1   B) −2

C) 10   D) No solution.

91) Find the value of 'x' that satisfies the below equation.

$- 39 = - 7 (- 3 - 7 x) - (- 6 x + 5)$

A) No solution.   B) −1

C) −15   D) 7

92) Find the value of 'a' that satisfies the below equation.

$- 34 = - 3 (6 + 6 a) + 4 (3 a - 4)$

A) 0   B) −10

C) −7   D) 1

93) Find the value of 'x' that satisfies the below equation.

$8 (1 - x) - 5 (x - 1) = 13$

A) 5   B) 0

C) 3   D) 15

94) Find the value of 'a' that satisfies the below equation.

$3 (- 2 a + 5) - (- 6 a + 1) = - 3$

A) No solution.   B) 8

C) −2   D) −16

95) Find the value of 'k' that satisfies the below equation.

$$-31 = 2(1 - 3k) - 3(k + 2)$$

A) No solution.   B) 12

C) 3   D) −15

96) Find the value of 'x' that satisfies the below equation.

$$4(x + 1) - 8(x - 7) + 4x = 60$$

A) −8
B) 3
C) All real numbers.
D) −7

97) Find the value of 'b' that satisfies the below equation.

$$-8(-5b + 1) + 8(b + 7) = 0$$

A) −1   B) 9

C) −8   D) −11

98) Find the value of 'p' that satisfies the below equation.

$$3 = 3(1 - 5p) - 5(5p - 8)$$

A) All real numbers.   B) 1

C) −6   D) 8

99) Find the value of 'v' that satisfies the below equation.

$$-8(v - 6) - 4(v - 2) = 20$$

A) 9   B) 3

C) 0   D) −8

100) Find the value of 'r' that satisfies the below equation.

$$-2(3 + 4r) + 5(1 - 6r) = -1$$

A) 1   B) −13

C) 0   D) 3

101) Find the value of 'x' that satisfies the below equation.

$$-49 = -3(x + 5) - 7(x + 2)$$

A) −2   B) 1

C) −15   D) 2

102) Find the value of 'k' that satisfies the below equation.

$$-4(-4k - 2) - 3(4k + 1) = -7$$

A) −5   B) −8

C) −15   D) −3

103) Find the value of 'x' that satisfies the below equation.

$$24 = -5(x + 8) - 6(7x + 5)$$

A) 6
B) All real numbers.
C) −2
D) 11

104) Find the value of 'a' that satisfies the below equation.

$$56 = -(8a + 7) + 5(-3a + 8)$$

A) −3   B) 6

C) 9   D) −1

105) Find the value of 'a' that satisfies the below equation.

$$-6(-7-5a)+5(-2a+1)=67$$

A) 9   B) 1

C) 14   D) 7

106) Find the value of 'm' that satisfies the below equation.

$$12=6(4m+7)-6(4m+5)$$

A) 14
B) All real numbers.
C) 10
D) −7

107) Find the value of 'a' that satisfies the below equation.

$$20=-4(1-6a)+4(4a+6)$$

A) −9   B) −16

C) 0   D) 13

108) Find the value of 'x' that satisfies the below equation.

$$7(3-8x)-(1-7x)=-29$$

A) All real numbers.
B) No solution.
C) 1
D) −1

109) Find the value of 'x' that satisfies the below equation.

$$15=-5(1+6x)-5(6+4x)$$

A) −9   B) −10

C) −1   D) 6

110) Find the value of 'k' that satisfies the below equation.

$$-4(1+7k)+6(2k-1)=-74$$

A) 4   B) 13

C) −9   D) −11

www.math-knots.com | www.a4ace.com

Algebra 1

1) Evaluate the below equation.

$$|x| = 19$$

   A) 14   B) 14, −14

   C) 19   D) 19, −19

2) Evaluate the below equation.

$$|x| = 24$$

   A) 19, −19   B) 19

   C) 1, −1   D) 24, −24

3) Evaluate the below equation.

$$|p| = 11$$

   A) 15, −15   B) 22, −22

   C) 17, −17   D) 11, −11

4) Evaluate the below equation.

$$|a| = 16$$

   A) 17, −17   B) 16, −16

   C) 35, −35   D) 32, −32

5) Evaluate the below equation.

$$|x| = 32$$

   A) 32, −32   B) 32

   C) 27   D) 27, −27

6) Evaluate the below equation.

$$|n| = 18$$

   A) 29, −29   B) 7

   C) 18, −18   D) 7, −7

7) Evaluate the below equation.

$$|x| = 4$$

   A) 26, −26   B) 4

   C) 4, −4   D) 33, −33

8) Evaluate the below equation.

$$|x| = 9$$

   A) 35, −35   B) 34

   C) 34, −34   D) 9, −9

9) Evaluate the below equation.

$$|p| = 38$$

   A) 28   B) 28, −28

   C) 24, −24   D) 38, −38

10) Evaluate the below equation.

$$|p| = 21$$

   A) 1, −1   B) 27, −27

   C) 26, −26   D) 21, −21

11) Evaluate the below equation.

$$|a + 1| = 4$$

A) 1, −1   B) 19, −7

C) 24, −24   D) 3, −5

12) Evaluate the below equation.

$$|-6x| = 6$$

A) −1, 1   B) 2, −2

C) 2   D) −1

13) Evaluate the below equation.

$$\left|\frac{r}{10}\right| = 5$$

A) 17, −3   B) 50, −50

C) 3, −13   D) 50

14) Evaluate the below equation.

$$|-5x| = 25$$

A) 7, −7   B) 6, 2

C) −5, 5   D) −5

15) Evaluate the below equation.

$$|r + 10| = 2$$

A) −8   B) −8, −12

C) 8, −8   D) 16, −10

16) Evaluate the below equation.

$$|x - 1| = 5$$

A) 18, −8   B) 9, −25

C) 6   D) 6, −4

17) Evaluate the below equation.

$$|x - 6| = 3$$

A) 9, 3   B) 8, −10

C) 2   D) 2, −10

18) Evaluate the below equation.

$$|a + 3| = 6$$

A) 3, −9   B) −5

C) 10, −10   D) −5, −7

19) Evaluate the below equation.

$$|-7 + r| = 11$$

A) 15, −3   B) 18, −4

C) −9, 9   D) 10, −12

20) Evaluate the below equation.

$$|b - 4| = 5$$

A) 6, −6   B) 8, −8

C) 5, −11   D) 9, −1

21) Evaluate the below equation.

$$\left|\frac{x}{3}\right| = 2$$

A) 4, −4   B) 6, −6

C) 4   D) 5, −7

22) Evaluate the below equation.

$$\left|\frac{x}{7}\right| = 1$$

A) 9, −11   B) 10, 8

C) 10   D) 7, −7

23) Evaluate the below equation.

$$\left|x - 1\right| = 1$$

A) 10, −10   B) 2, 0

C) 6, −8   D) −6, −8

24) Evaluate the below equation.

$$\left|x - 2\right| = 9$$

A) −5, −11   B) 11, −7

C) 6, 0   D) 6

25) Evaluate the below equation.

$$\left|9 + r\right| = 7$$

A) 2, −10   B) −2

C) −3, 3   D) −2, −16

26) Evaluate the below equation.

$$\left|7 + a\right| = 1$$

A) −6, −8   B) 5

C) 5, −5   D) −6

27) Evaluate the below equation.

$$\left|k + 7\right| = 2$$

A) 18, −18   B) 12, −12

C) −5, −9   D) 9, 5

28) Evaluate the below equation.

$$\left|6a\right| = 12$$

A) −8, −10   B) 2

C) 2, −2   D) 8, −8

29) Evaluate the below equation.

$$\left|n + 10\right| = 10$$

A) 0, −20   B) 5, −25

C) −7, 7   D) 1, −1

30) Evaluate the below equation.

$$\left|-5x\right| = 45$$

A) 14   B) 8, −10

C) −9, 9   D) 14, −14

31) Evaluate the below equation.

$$|n + 10| = 9$$

A) $-5, 5$ B) $18, -18$

C) $-1, -19$ D) $-8, 8$

32) Evaluate the below equation.

$$|-8r| = 48$$

A) $-6, 6$ B) -6

C) $5, -7$ D) $11, 3$

33) Evaluate the below equation.

$$|10b| = 90$$

A) 9 B) $9, -9$

C) $2, -16$ D) 2

34) Evaluate the below equation.

$$|-6 + p| = 4$$

A) $10, 2$ B) $18, -8$

C) $-10, 10$ D) $10, -20$

35) Evaluate the below equation.

$$|x - 2| = 6$$

A) $-2, 2$ B) $6, -6$

C) $8, -4$ D) $12, -12$

36) Evaluate the below equation.

$$|-9x + 1| = 19$$

A) 2 B) $2, -\dfrac{8}{5}$

C) $14, -6$ D) $-2, \dfrac{20}{9}$

37) Evaluate the below equation.

$$|-10n - 3| = 77$$

A) $-8, \dfrac{37}{5}$ B) $-\dfrac{20}{3}, 8$

C) $3, -\dfrac{15}{4}$ D) $\dfrac{7}{2}, -6$

38) Evaluate the below equation.

$$|6 + 6x| = 36$$

A) $5, -7$ B) 7

C) $7, -\dfrac{47}{5}$ D) $14, -10$

39) Evaluate the below equation.

$$|8 + 5x| = 22$$

A) $-10, -6$ B) $7, -\dfrac{39}{7}$

C) $\dfrac{14}{5}, -6$ D) $-4, \dfrac{11}{5}$

40) Evaluate the below equation.

$$|10p - 6| = 46$$

A) $-1, 13$ B) $-4, \dfrac{38}{5}$

C) -1 D) $\dfrac{26}{5}, -4$

www.math-knots.com | www.a4ace.com

41) Evaluate the below equation.

$$\left|-5n - 9\right| = 4$$

A) $5, -\dfrac{20}{3}$ B) $-6, 7$

C) $-\dfrac{8}{3}, 5$ D) $-\dfrac{13}{5}, -1$

42) Evaluate the below equation.

$$\left|3x - 2\right| = 8$$

A) $-9, 17$ B) 4

C) $\dfrac{10}{3}, -2$ D) $4, -\dfrac{22}{5}$

43) Evaluate the below equation.

$$\left|x - 10\right| = 7$$

A) $17, 3$ B) $\dfrac{21}{2}, -9$

C) 1 D) $1, -\dfrac{14}{5}$

44) Evaluate the below equation.

$$\left|-9 + 8b\right| = 39$$

A) $6, -\dfrac{15}{4}$ B) $6, -26$

C) $-10, \dfrac{58}{7}$ D) $4, -6$

45) Evaluate the below equation.

$$\left|3 - 7k\right| = 66$$

A) $6, -5$ B) $-9, \dfrac{69}{7}$

C) 6 D) -9

46) Evaluate the below equation.

$$\left|-5x + 9\right| = 11$$

A) 0 B) $-\dfrac{2}{5}$

C) $-\dfrac{2}{5}, 4$ D) $0, -\dfrac{14}{3}$

47) Evaluate the below equation.

$$\left|6k - 9\right| = 27$$

A) $3, -2$ B) 2

C) $6, -3$ D) $2, -6$

48) Evaluate the below equation.

$$\left|2b - 9\right| = 3$$

A) 6 B) $8, -\dfrac{29}{4}$

C) $0, -3$ D) $6, 3$

49) Evaluate the below equation.

$$\left|7 + 4x\right| = 33$$

A) $\dfrac{9}{4}, -1$ B) $7, -\dfrac{13}{2}$

C) $\dfrac{13}{2}, -10$ D) $\dfrac{16}{3}, -8$

50) Evaluate the below equation.

$$\left|6 + 9x\right| = 33$$

A) $22, -10$ B) $3, -\dfrac{13}{3}$

C) $6, -4$ D) 22

51) Evaluate the below equation.

$$|3n - 5| = 7$$

A) $10, -19$ B) $4, -\dfrac{2}{3}$

C) $8, -18$ D) $\dfrac{3}{4}, -1$

52) Evaluate the below equation.

$$|3v + 2| = 1$$

A) $-\dfrac{1}{3}, -1$ B) $5, -\dfrac{49}{9}$

C) $-\dfrac{1}{3}$ D) $\dfrac{20}{3}, -6$

53) Evaluate the below equation.

$$|6 + 9x| = 69$$

A) 7 B) $7, -\dfrac{25}{3}$

C) $1, \dfrac{1}{2}$ D) $-1, \dfrac{25}{7}$

54) Evaluate the below equation.

$$|9v + 3| = 51$$

A) $9, -\dfrac{31}{3}$ B) $\dfrac{16}{3}, -6$

C) $-\dfrac{32}{3}, 10$ D) $-7, \dfrac{37}{5}$

55) Evaluate the below equation.

$$|4k + 4| = 32$$

A) $7, -9$ B) $6, -\dfrac{29}{4}$

C) $\dfrac{26}{3}, -2$ D) $-\dfrac{1}{2}, -2$

56) Evaluate the below equation.

$$|5 + 5v| = 35$$

A) $6, -8$ B) 6

C) $-\dfrac{69}{7}, 7$ D) $-17, 1$

57) Evaluate the below equation.

$$|10x + 7| = 93$$

A) $-6, \dfrac{28}{3}$ B) $\dfrac{43}{5}, -10$

C) $3, \dfrac{3}{2}$ D) $-\dfrac{4}{5}, 1$

58) Evaluate the below equation.

$$|6n - 1| = 49$$

A) $\dfrac{25}{3}, -8$ B) $-2, \dfrac{15}{4}$

C) $5, -\dfrac{7}{2}$ D) $-\dfrac{33}{7}, 3$

59) Evaluate the below equation.

$$|2 + 3b| = 1$$

A) $-\dfrac{1}{3}, -1$ B) $-\dfrac{6}{5}, 4$

C) $8, 0$ D) $-\dfrac{1}{3}$

60) Evaluate the below equation.

$$|-5 - 7x| = 61$$

A) $-\dfrac{49}{9}, 5$ B) $\dfrac{71}{9}, -7$

C) $7, -15$ D) $-\dfrac{66}{7}, 8$

61) Evaluate the below equation.

$$|3x + 2| = 20$$

A) $6, -\dfrac{22}{3}$ B) $4, -\dfrac{20}{7}$

C) $\dfrac{13}{3}, -5$ D) $3, -4$

62) Evaluate the below equation.

$$9|n - 9| = 18$$

A) $11, 7$ B) $9, -3$

C) $38, -52$ D) $-8, 8$

63) Evaluate the below equation.

$$3 + \left|\dfrac{x}{7}\right| = 4$$

A) $2, -10$ B) 7

C) $0, -14$ D) $7, -7$

64) Evaluate the below equation.

$$|-6x| + 6 = 66$$

A) $-5, 5$ B) -10

C) -5 D) $-10, 10$

65) Evaluate the below equation.

$$-7 + |4p| = 33$$

A) $9, 7$ B) $15, -21$

C) $6, -4$ D) $10, -10$

66) Evaluate the below equation.

$$\left|\dfrac{p}{5}\right| - 1 = 0$$

A) -4 B) $5, -5$

C) $9, -9$ D) $-4, 4$

67) Evaluate the below equation.

$$6 + \left|\dfrac{r}{3}\right| = 8$$

A) $7, -3$ B) $14, 4$

C) $6, -6$ D) 6

68) Evaluate the below equation.

$$|7 + x| + 4 = 9$$

A) $9, -9$ B) $-2, -12$

C) 8 D) $8, -28$

69) Evaluate the below equation.

$$\dfrac{|6x|}{8} = 5$$

A) $-4, 4$ B) $2, -18$

C) $\dfrac{20}{3}$ D) $\dfrac{20}{3}, -\dfrac{20}{3}$

70) Evaluate the below equation.

$$|-3b| - 5 = 7$$

A) 4 B) $18, -4$

C) $-4, 4$ D) $4, -20$

71) Evaluate the below equation.

$$4\left|\frac{x}{2}\right| = 20$$

A) 10 , −10   B) 1 , −7

C) −6 , −14   D) 1 , −11

72) Evaluate the below equation.

$$\frac{|10\,a|}{5} = 2$$

A) 20 , −4   B) 22 , −6

C) 6 , −6   D) 1 , −1

73) Evaluate the below equation.

$$|7v| + 2 = 30$$

A) 7 , −7   B) −4 , −6

C) 15 , −7   D) 4 , −4

74) Evaluate the below equation.

$$7|b + 2| = 63$$

A) 8 , −8   B) 7 , −11

C) 4 , 0   D) −3 , 3

75) Evaluate the below equation.

$$\left|\frac{m}{7}\right| - 10 = -9$$

A) 6 , 2   B) 10 , −10

C) 7 , −7   D) 7

76) Evaluate the below equation.

$$-7|9n| = -63$$

A) 1 , −1   B) 10

C) 23 , −3   D) 10 , −10

77) Evaluate the below equation.

$$|x - 7| - 7 = -1$$

A) $-\frac{6}{5}, \frac{6}{5}$ B) 13 , 1

C) 0 , −4   D) 2 , −10

78) Evaluate the below equation.

$$|n + 2| + 6 = 13$$

A) 5 , −9   B) −2 , −4

C) $-\frac{12}{7}, \frac{12}{7}$ D) −9 , 9

79) Evaluate the below equation.

$$-7|-7x| = -49$$

A) −1 , 1   B) 9

C) 9 , −1   D) 31 , −49

80) Evaluate the below equation.

$$10 + |6b| = 64$$

A) 8 , −28   B) 60 , −40

C) 10 , −10   D) 9 , −9

81) Evaluate the below equation.

$$-3 + \left| \frac{x}{10} \right| = -2$$

A) $10, -10$ B) $23, -17$

C) 18 D) $18, -6$

82) Evaluate the below equation.

$$7 \left| x - 3 \right| = 42$$

A) $18, -4$ B) $4, -4$

C) $9, -3$ D) 4

83) Evaluate the below equation.

$$-5 + \left| \frac{x}{10} \right| = -4$$

A) $10, -10$ B) $17, 1$

C) $-14, 14$ D) 17

84) Evaluate the below equation.

$$-3 + \left| p - 4 \right| = 0$$

A) $7, 1$ B) $40, -20$

C) $5, -5$ D) $-1, 1$

85) Evaluate the below equation.

$$\frac{\left| n + 6 \right|}{5} = 5$$

A) $3, -3$ B) $19, -31$

C) $6, -20$ D) $17, -31$

86) Evaluate the below equation.

$$\frac{\left| -2b \right|}{6} = 2$$

A) $-\frac{9}{4}, \frac{9}{4}$ B) -6

C) $-6, 6$ D) $-1, -11$

    99

Algebra 1

1) Solve the equation for the indicated variable.

$$x - k = v - w \text{ , for } x$$

A) $x = -k - v - w$

B) $x = -k - v + w$

C) $x = k + v - w$

D) $x = k - v - w$

2) Solve the equation for the indicated variable.

$$a\,c = d\,r \text{ , for } a$$

A) $a = \dfrac{dr}{c}$ B) $a = -dr + c$

C) $a = drc$ D) $a = -\dfrac{dr}{c}$

3) Solve the equation for the indicated variable.

$$\dfrac{c}{x} = d\,r, \text{ for } x$$

A) $x = -\dfrac{c}{dr}$ B) $x = -cdr$

C) $x = \dfrac{c}{dr}$ D) $x = \dfrac{dr}{c}$

4) Solve the equation for the indicated variable.

$$kx = \dfrac{v}{w} \text{ , for } x$$

A) $x = vkw$ B) $x = -\dfrac{v}{kw}$

C) $x = \dfrac{v}{kw}$ D) $x = v + kw$

5) Solve the equation for the indicated variable.

$$z = \dfrac{am}{b} \text{ , for } a$$

A) $a = zb + m$ B) $a = -zbm$

C) $a = zbm$ D) $a = \dfrac{zb}{m}$

6) Solve the equation for the indicated variable.

$$am = n + p \text{ , for } a$$

A) $a = \dfrac{n + p}{m}$ B) $a = \dfrac{-n - p}{m}$

C) $a = \dfrac{m}{n + p}$ D) $a = -\dfrac{m}{n + p}$

7) Solve the equation for the indicated variable.

$$z = -y + \frac{m}{x} \text{, for } x$$

A) $x = mz - my$

B) $x = \frac{m}{-z + y}$

C) $x = \frac{m}{z + y}$

D) $x = \frac{m}{-z - y}$

8) Solve the equation for the indicated variable.

$$z = \frac{xy}{m} \text{, for } x$$

A) $x = -zmy$

B) $x = \frac{zm}{y}$

C) $x = zm + y$

D) $x = -\frac{y}{zm}$

9) Solve the equation for the indicated variable.

$$u = \frac{b}{ak} \text{, for } a$$

A) $a = \frac{b}{uk}$

B) $a = -buk$

C) $a = \frac{uk}{b}$

D) $a = buk$

10) Solve the equation for the indicated variable.

$$cx = \frac{d}{r} \text{, for } x$$

A) $x = \frac{d}{cr}$

B) $x = -\frac{cr}{d}$

C) $x = -d + cr$

D) $x = dcr$

11) Solve the equation for the indicated variable.

$$u = ak - b \text{, for } a$$

A) $a = u + b + k$

B) $a = u + b - k$

C) $a = \frac{u + b}{k}$

D) $a = \frac{k}{u - b}$

12) Solve the equation for the indicated variable.

$$\frac{k}{x} = v - w \text{, for } x$$

A) $x = k - v - w$

B) $x = \frac{-v + w}{k}$

C) $x = \frac{k}{v - w}$

D) $x = \frac{v - w}{k}$

www.math-knots.com | www.a4ace.com

13) Solve the equation for the indicated variable.

$$g = c\,a - b \text{ , for } a$$

A) $a = \dfrac{g+b}{c}$

B) $a = \dfrac{-g-b}{c}$

C) $a = -c\,(g+b)$

D) $a = c\,g + c\,b$

14) Solve the equation for the indicated variable.

$$u = \dfrac{xy}{k} \text{ , for } x$$

A) $x = \dfrac{uk}{y}$ B) $x = -\dfrac{uk}{y}$

C) $x = -\dfrac{y}{uk}$ D) $x = uk + y$

15) Solve the equation for the indicated variable.

$$m + x = n + p \text{ , for } x$$

A) $x = -m - n + p$

B) $x = m + n + p$

C) $x = m - n + p$

D) $x = -m + n + p$

16) Solve the equation for the indicated variable.

$$g = a\,c + b \text{ , for } a$$

A) $a = g + b + c$

B) $a = \dfrac{g-b}{c}$

C) $a = -c\,g + c\,b$

D) $a = -g - b + c$

17) Solve the equation for the indicated variable.

$$\dfrac{x}{c} = \dfrac{d}{r} \text{ , for } x$$

A) $x = cd\,r$ B) $x = \dfrac{cd}{r}$

C) $x = \dfrac{r}{cd}$ D) $x = -\dfrac{cd}{r}$

18) Solve the equation for the indicated variable.

$$x + k = v - w \text{ , for } x$$

A) $x = -k + v - w$

B) $x = -k - v + w$

C) $x = -k + v + w$

D) $x = -k - v - w$

www.math-knots.com | www.a4ace.com

19) Solve the equation for the indicated variable.

$$a - m = n + p \text{ , for } a$$

A) $a = -p - n + m$

B) $a = -m - n - p$

C) $a = m + p - n$

D) $a = m + n + p$

20) Solve the equation for the indicated variable.

$$k + a = w + v \text{ , for } a$$

A) $a = -k + w - v$

B) $a = k - w + v$

C) $a = -k + w + v$

D) $a = k - v + w$

21) Solve the equation for the indicated variable.

$$u = kx + yx \text{ , for } x$$

A) $x = -uk + uy$ B) $x = \dfrac{-k - y}{u}$

C) $x = \dfrac{u}{k + y}$ D) $x = \dfrac{k + y}{u}$

22) Solve the equation for the indicated variable.

$$ua = \dfrac{a + b}{k} \text{ , for } a$$

A) $a = -\dfrac{b}{uk + 1}$

B) $a = \dfrac{b}{-uk - 1}$

C) $a = -buk + b$

D) $a = \dfrac{b}{uk - 1}$

23) Solve the equation for the indicated variable.

$$zma = a + b \text{ , for } a$$

A) $a = \dfrac{b}{zm - 1}$

B) $a = bzm + b$

C) $a = -bzm + b$

D) $a = -\dfrac{b}{-zm - 1}$

24) Solve the equation for the indicated variable.

$$u = kx + yx \text{ , for } x$$

A) $x = \dfrac{k + y}{u}$ B) $x = -\dfrac{u}{k + y}$

C) $x = -\dfrac{u}{-k - y}$ D) $x = \dfrac{u}{k + y}$

25) Solve the equation for the indicated variable.

$$z + m\,a = b\,a \text{ , for } a$$

A) $a = -\dfrac{z}{-m+b}$

B) $a = zm - zb$

C) $a = \dfrac{z}{-m+b}$

D) $a = -\dfrac{z}{-m-b}$

26) Solve the equation for the indicated variable.

$$u + k\,x = y\,x \text{ , for } x$$

A) $x = \dfrac{k-y}{u}$

B) $x = -\dfrac{u}{k+y}$

C) $x = \dfrac{u}{-k+y}$

D) $x = -\dfrac{u}{-k+y}$

27) Solve the equation for the indicated variable.

$$u = \dfrac{x+y}{kx} \text{ , for } x$$

A) $x = -\dfrac{y}{uk+1}$

B) $x = -\dfrac{y}{-uk+1}$

C) $x = \dfrac{y}{uk-1}$

D) $x = \dfrac{uk+1}{y}$

28) Solve the equation for the indicated variable.

$$z + m\,x = y\,x \text{ , for } x$$

A) $x = -\dfrac{z}{m+y}$

B) $x = \dfrac{m-y}{z}$

C) $x = \dfrac{z}{-m+y}$

D) $x = z - m + y$

29) Solve the equation for the indicated variable.

$$u + k\,x = y\,x \text{ , for } x$$

A) $x = \dfrac{u}{-k-y}$

B) $x = uk - uy$

C) $x = \dfrac{u}{-k+y}$

D) $x = \dfrac{-k-y}{u}$

30) Solve the equation for the indicated variable.

$$g = \dfrac{x+y}{cx} \text{ , for } x$$

A) $x = -y\,g\,c - y$

B) $x = \dfrac{y}{gc-1}$

C) $x = -y + g\,c - 1$

D) $x = \dfrac{-g\,c-1}{y}$

31) Solve the equation for the indicated variable.

$$za = \frac{a+b}{m}, \text{ for } a$$

A) $a = -\dfrac{b}{zm+1}$

B) $a = \dfrac{b}{zm-1}$

C) $a = \dfrac{-zm-1}{b}$

D) $a = -\dfrac{b}{-zm-1}$

32) Solve the equation for the indicated variable.

$$z\,m\,x = x + y, \text{ for } x$$

A) $x = \dfrac{y}{zm-1}$

B) $x = \dfrac{-zm+1}{y}$

C) $x = y + zm - 1$

D) $x = \dfrac{zm-1}{y}$

33) Solve the equation for the indicated variable.

$$zx = \frac{x+y}{m}, \text{ for } x$$

A) $x = -yzm + y$

B) $x = \dfrac{y}{zm-1}$

C) $x = -\dfrac{y}{-zm-1}$

D) $x = \dfrac{-zm-1}{y}$

34) Solve the equation for the indicated variable.

$$g = cx + yx, \text{ for } x$$

A) $x = -\dfrac{g}{c-y}$

B) $x = \dfrac{g}{c+y}$

C) $x = g + c + y$

D) $x = g + c - y$

35) Solve the equation for the indicated variable.

$$u + kx = yx, \text{ for } x$$

A) $x = \dfrac{u}{k+y}$ B) $x = \dfrac{u}{-k+y}$

C) $x = \dfrac{-k+y}{u}$ D) $x = -uk + uy$

36) Solve the equation for the indicated variable.

$$ux = \frac{x+y}{k}, \text{ for } x$$

A) $x = \dfrac{uk-1}{y}$ B) $x = \dfrac{y}{-uk+1}$

C) $x = \dfrac{y}{uk-1}$ D) $x = \dfrac{-uk+1}{y}$

37) Solve the equation for the indicated variable.

$$g + c\,a = b\,a \text{ , for } a$$

A) $a = g + c + b$

B) $a = -g - c + b$

C) $a = \dfrac{g}{-c + b}$

D) $a = -g\,c + g\,b$

38) Solve the equation for the indicated variable.

$$u = k\,x + y\,x \text{ , for } x$$

A) $x = u - k + y$

B) $x = \dfrac{u}{-k + y}$

C) $x = \dfrac{u}{k + y}$

D) $x = \dfrac{k + y}{u}$

39) Solve the equation for the indicated variable.

$$z\,m\,a = a + b \text{ , for } a$$

A) $a = \dfrac{-zm + 1}{b}$

B) $a = \dfrac{b}{zm - 1}$

C) $a = -\dfrac{b}{-zm - 1}$

D) $a = \dfrac{zm + 1}{b}$

40) Solve the equation for the indicated variable.

$$g\,c\,x = x + y \text{ , for } x$$

A) $x = -y + g\,c - 1$

B) $x = y + g\,c - 1$

C) $x = \dfrac{-gc + 1}{y}$

D) $x = \dfrac{y}{gc - 1}$

41) Solve the equation for the indicated variable.

$$x\,c = y + d - r \text{ , for } x$$

A) $x = \dfrac{y - d - r}{c}$

B) $x = \dfrac{c}{y + d - r}$

C) $x = c\,(\,y + d - r\,)$

D) $x = \dfrac{y + d - r}{c}$

42) Solve the equation for the indicated variable.

$$g = \dfrac{cxr + d}{r} \text{ , for } x$$

A) $x = \dfrac{-gr + d}{cr}$

B) $x = \dfrac{gr - d}{cr}$

C) $x = \dfrac{cr}{gr - d}$

D) $x = -g\,r + d + c\,r$

43) Solve the equation for the indicated variable.

$$xc = \frac{r}{d} + y \text{ , for } x$$

A) $x = yd - r + cd$

B) $x = yd + r + cd$

C) $x = \dfrac{cd}{yd + r}$

D) $x = \dfrac{yd + r}{cd}$

44) Solve the equation for the indicated variable.

$$\frac{k}{a} = b(w - v) \text{ , for } a$$

A) $a = kb(w - v)$

B) $a = \dfrac{bw + bv}{k}$

C) $a = \dfrac{k}{b(w - v)}$

D) $a = -kbw + kbv$

45) Solve the equation for the indicated variable.

$$ak = b + v + w \text{ , for } a$$

A) $a = \dfrac{k}{b - v - w}$

B) $a = \dfrac{b + v + w}{k}$

C) $a = \dfrac{b - v + w}{k}$

D) $a = k(b - v + w)$

46) Solve the equation for the indicated variable.

$$\frac{k}{a} = \frac{w + v}{b} \text{ , for } a$$

A) $a = \dfrac{kb}{w + v}$

B) $a = -kb + w + v$

C) $a = \dfrac{kb}{-w - v}$

D) $a = \dfrac{w + v}{kb}$

47) Solve the equation for the indicated variable.

$$ka = b + w - v \text{ , for } a$$

A) $a = \dfrac{b + w - v}{k}$

B) $a = -b + w + v + k$

C) $a = k(b - w + v)$

D) $a = b + w - v + k$

48) Solve the equation for the indicated variable.

$$kx = y(v - w) \text{ , for } x$$

A) $x = \dfrac{k}{yv + yw}$

B) $x = -\dfrac{k}{yv - yw}$

C) $x = yv - yw - k$

D) $x = \dfrac{y(v - w)}{k}$

49) Solve the equation for the indicated variable.

$$x\,m = n - p + y \text{ , for } x$$

A) $x = \dfrac{n - p + y}{m}$

B) $x = \dfrac{-n - p + y}{m}$

C) $x = \dfrac{n + p - y}{m}$

D) $x = \dfrac{-n + p + y}{m}$

50) Solve the equation for the indicated variable.

$$c + x = y\,(\,d + r\,) \text{ , for } x$$

A) $x = y\,d + y\,r + c$

B) $x = y\,d + y\,r - c$

C) $x = -\,y\,d - y\,r + c$

D) $x = -\,y\,r + c + y\,d$

51) Solve the equation for the indicated variable.

$$\dfrac{m}{a} = npb \text{ , for } a$$

A) $a = \dfrac{npb}{m}$

B) $a = \dfrac{m}{npb}$

C) $a = m + npb$

D) $a = -\dfrac{npb}{m}$

52) Solve the equation for the indicated variable.

$$g = (\,c + x\,)\,(\,d + r\,) \text{ , for } x$$

A) $x = \dfrac{d - r}{g - cd - cr}$

B) $x = \dfrac{-g + cd + cr}{-d + r}$

C) $x = \dfrac{g - cd - cr}{d + r}$

D) $x = g - c\,d - c\,r - d + r$

53) Solve the equation for the indicated variable.

$$m + a = b\,(\,p + n\,) \text{ , for } a$$

A) $a = -\,b\,p - b\,n + m$

B) $a = b\,p + b\,n - m$

C) $a = -\,b\,p + b\,n - m$

D) $a = b\,p + b\,n + m$

54) Solve the equation for the indicated variable.

$$z = \dfrac{x + m}{n - p} \text{ , for } x$$

A) $x = -\,z\,n - z\,p + m$

B) $x = z\,n - z\,p + m$

C) $x = z\,n - z\,p - m$

D) $x = z\,n + z\,p + m$

 **Algebra 1**

55) Solve the equation for the indicated variable.

$$z = \frac{am}{n+p} \text{ , for } a$$

A) $a = \dfrac{z(n+p)}{m}$

B) $a = -\dfrac{m}{zn-zp}$

C) $a = zn + zp - m$

D) $a = \dfrac{-zn - zp}{m}$

56) Solve the equation for the indicated variable.

$$kx = wv + y \text{ , for } x$$

A) $x = kwv + ky$

B) $x = -\dfrac{k}{wv + y}$

C) $x = \dfrac{wv + y}{k}$

D) $x = \dfrac{k}{wv + y}$

57) Solve the equation for the indicated variable.

$$\frac{c}{a} = b - dr \text{ , for } a$$

A) $a = cdr + cb$

B) $a = \dfrac{c}{-dr + b}$

C) $a = \dfrac{-dr + b}{c}$

D) $a = \dfrac{-dr - b}{c}$

58) Solve the equation for the indicated variable.

$$u = \frac{w - v}{x + k} \text{ , for } x$$

A) $x = uk + w - v + u$

B) $x = \dfrac{-uk + w - v}{u}$

C) $x = \dfrac{uk - w - v}{u}$

D) $x = -uk + w - v - u$

59) Solve the equation for the indicated variable.

$$x + c = \frac{y}{d + r} \text{ , for } x$$

A) $x = (cd + cr + y)(d + r)$

B) $x = (cd - cr - y)(d + r)$

C) $x = \dfrac{-cd - cr + y}{d + r}$

D) $x = \dfrac{d + r}{-cd - cr + y}$

60) Solve the equation for the indicated variable.

$$x - k = w + v - y \text{ , for } x$$

A) $x = -w + v + k + y$

B) $x = k + w + v - y$

C) $x = w - k - v - y$

D) $x = -k - w + v + y$

61) Solve the equation for the indicated variable.

$$x\,m = p - n + y\,x \text{, for } x$$

A) $x = \dfrac{-p - n}{-m - y}$

B) $x = \dfrac{p - n}{m - y}$

C) $x = \dfrac{-p + n}{m - y}$

D) $x = p - n + m - y$

62) Solve the equation for the indicated variable.

$$c + a = r + d + b\,a \text{, for } a$$

A) $a = (c - r - d)(b - 1)$

B) $a = (c - r - d)(b + 1)$

C) $a = \dfrac{-c - r - d}{b - 1}$

D) $a = \dfrac{c - r - d}{b - 1}$

63) Solve the equation for the indicated variable.

$$c + a = d\,r\,b\,a \text{, for } a$$

A) $a = \dfrac{-d\,r\,b - 1}{c}$

B) $a = \dfrac{c}{d\,r\,b - 1}$

C) $a = c + d\,r\,b + 1$

D) $a = \dfrac{c}{-d\,r\,b + 1}$

64) Solve the equation for the indicated variable.

$$z = \dfrac{x\,m + p}{x\,n} \text{, for } x$$

A) $x = \dfrac{-z\,n + m}{p}$

B) $x = \dfrac{p}{z\,n - m}$

C) $x = \dfrac{z\,n - m}{p}$

D) $x = -p\,z\,n - p\,m$

65) Solve the equation for the indicated variable.

$$c - x = r\,d + y\,x \text{, for } x$$

A) $x = (c - r\,d)(-y + 1)$

B) $x = \dfrac{c - r\,d}{y + 1}$

C) $x = \dfrac{-y - 1}{c - r\,d}$

D) $x = (c - r\,d)(y + 1)$

66) Solve the equation for the indicated variable.

$$a - c = r + d + b\,a \text{, for } a$$

A) $a = \dfrac{c - r + d}{b + 1}$

B) $a = \dfrac{-b + 1}{-c + r - d}$

C) $a = \dfrac{c + r + d}{-b + 1}$

D) $a = c + r + d + b + 1$

67) Solve the equation for the indicated variable.

$$k + x = v + w + y\,x \text{, for } x$$

A) $x = (-k - v + w)(y - 1)$

B) $x = \dfrac{-y - 1}{k - v - w}$

C) $x = \dfrac{y^- 1}{k + v + w}$

D) $x = \dfrac{k - v - w}{y - 1}$

68) Solve the equation for the indicated variable.

$$k\,a = v - w + b\,a \text{, for } a$$

A) $a = -v + w + k - b$

B) $a = \dfrac{v - w}{k - b}$

C) $a = \dfrac{k - b}{v + w}$

D) $a = (v - w)(k + b)$

69) Solve the equation for the indicated variable.

$$a\,m = n - p + b\,a \text{, for } a$$

A) $a = n - p + m - b$

B) $a = \dfrac{n - p}{m - b}$

C) $a = \dfrac{n + p}{-m - b}$

D) $a = (-n + p)(m - b)$

70) Solve the equation for the indicated variable.

$$g = a\,c + a\,d + a\,r \text{, for } a$$

A) $a = \dfrac{-c - d - r}{g}$

B) $a = \dfrac{g}{c + d + r}$

C) $a = g(-c + d - r)$

D) $a = \dfrac{c - d - r}{g}$

71) Solve the equation for the indicated variable.

$$z = x\,m + x\,n + p \text{, for } x$$

A) $x = -z - p + m + n$

B) $x = \dfrac{z - p}{-m - n}$

C) $x = \dfrac{z - p}{m + n}$

D) $x = \dfrac{-z - p}{m + n}$

72) Solve the equation for the indicated variable.

$$g = \dfrac{xc + r}{xd} \text{, for } x$$

A) $x = -r\,g\,d + r\,c$

B) $x = \dfrac{r}{gd - c}$

C) $x = r + g\,d + c$

D) $x = -\dfrac{r}{gd - c}$

73) Solve the equation for the indicated variable.

$$a - m = p\,n - b\,a, \text{ for } a$$

A) $a = m + pn + b + 1$

B) $a = \dfrac{m + pn}{b + 1}$

C) $a = \dfrac{-m + pn}{b - 1}$

D) $a = \dfrac{-b - 1}{m + pn}$

74) Solve the equation for the indicated variable.

$$m - x = y\,x + p\,n, \text{ for } x$$

A) $x = \dfrac{y + 1}{-m - pn}$

B) $x = \dfrac{-m + pn}{-y + 1}$

C) $x = (-m - pn)(y + 1)$

D) $x = \dfrac{m - pn}{y + 1}$

75) Solve the equation for the indicated variable.

$$g = x\,c + x\,d + r, \text{ for } x$$

A) $x = (-g - r)(c + d)$

B) $x = \dfrac{-c - d}{g - r}$

C) $x = \dfrac{g - r}{c + d}$

D) $x = g - r + c + d$

76) Solve the equation for the indicated variable.

$$a - c = r - d + b\,a, \text{ for } a$$

A) $a = \dfrac{c + r - d}{-b + 1}$

B) $a = \dfrac{b + 1}{c + r - d}$

C) $a = (-c - r - d)(-b + 1)$

D) $a = -c - r + d - b + 1$

Answer the questions 77 - 83 based on the below data.

| 6 | 7 | 6 | 7 | 4 | 7 | 4 | 4 |
| 6 | | | | | | | |

77) Find the median

78) Find the mean

79) Find the lower quartile (Q$_1$)

80) Find the upper quartile (Q$_3$)

81) Find the inter quartile range

82) Find the standard deviation (*s)*

83) Find the population standard deviation (σ)

Answer the questions 84 - 90 based on the below data.

| 45 | 41 | 46 | 43 | 43 | 43 | 45 |
| 48 | 54 | 38 | | | | |

84) Find the median

85) Find the mean

86) Find the lower quartile (Q$_1$)

87) Find the upper quartile (Q$_3$)

88) Find the inter quartile range

89) Find the standard deviation (*s)*

90) Find the population standard deviation (σ)

Answer the questions 91 - 97 based on the below data.

12.4	66.4	20.8	55	33.6	12.6
30.8	26.4	23	51	19.6	

91) Find the median

92) Find the mean

93) Find the lower quartile (Q_1)

94) Find the upper quartile (Q_3)

95) Find the inter quartile range

96) Find the standard deviation (*s*)

97) Find the population standard deviation (σ)

Answer the questions 98 - 104 based on the below data.

20	13	18	16	17	17	17
21	16	13				

98) Find the median

99) Find the mean

100) Find the lower quartile (Q_1)

101) Find the upper quartile (Q_3)

102) Find the inter quartile range

103) Find the standard deviation (*s*)

104) Find the population standard deviation (σ)

Answer the questions 105 - 111 based on the below data.

| 3 | 1 | 2 | 2 | 2 | 4 | 6 | 2 |
| 2 | 2 | | | | | | |

105) Find the median

106) Find the mean

107) Find the lower quartile (Q_1)

108) Find the upper quartile (Q_3)

109) Find the inter quartile range

110) Find the standard deviation (*s*)

111) Find the population standard deviation (σ)

Answer the questions 112 - 118 based on the below data.

29.5	26.1	40.6	39.2	24.7
27.9	27.2	32	38.9	29.6
28.4				

112) Find the median

113) Find the mean

114) Find the lower quartile (Q_1)

115) Find the upper quartile (Q_3)

116) Find the inter quartile range

117) Find the standard deviation (*s*)

118) Find the population standard deviation (σ)

     www.math-knots.com | www.a4ace.com

Answer the questions 119 - 125 based on the below data.

44	46	32	43	51	37	45
50	53	46	56			

119) Find the median

120) Find the mean

121) Find the lower quartile (Q_1)

122) Find the upper quartile (Q_3)

123) Find the inter quartile range

124) Find the standard deviation (*s*)

125) Find the population standard deviation (σ)

Answer the questions 126 - 132 based on the below data.

48	51	54	47	53	36	39
49	55	40				

126) Find the median

127) Find the mean

128) Find the lower quartile (Q_1)

129) Find the upper quartile (Q_3)

130) Find the inter quartile range

131) Find the standard deviation (*s*)

132) Find the population standard deviation (σ)

www.math-knots.com | www.a4ace.com

Answer the questions 133 - 139 based on the below data.

13	14	15	17	12	15	16
17	16	12	15			

133) Find the median

134) Find the mean

135) Find the lower quartile (Q_1)

136) Find the upper quartile (Q_3)

137) Find the inter quartile range

138) Find the standard deviation (*s*)

139) Find the population standard
    deviation (σ)

Answer the questions 140 - 146 based on the below data.

7	5	5	8	20	6	4	8
15							

140) Find the median

141) Find the mean

142) Find the lower quartile (Q_1)

143) Find the upper quartile (Q_3)

144) Find the inter quartile range

145) Find the standard deviation (*s*)

146) Find the population standard
    deviation (σ)

Algebra 1

Answer the questions 147 - 153 based on the below data.

29	23	33.8	32.6	38	37.5
30	27.6	35.8	44.5		

147) Find the median

148) Find the mean

149) Find the lower quartile (Q_1)

150) Find the upper quartile (Q_3)

151) Find the inter quartile range

152) Find the standard deviation (*s*)

153) Find the population standard deviation (σ)

Answer the questions 154 - 160 based on the below data.

17	15	12	18	18	17	16
17	13	16				

154) Find the median

155) Find the mean

156) Find the lower quartile (Q_1)

157) Find the upper quartile (Q_3)

158) Find the inter quartile range

159) Find the standard deviation (*s*)

160) Find the population standard deviation (σ)

1) Express the below as algebraic expression or equation or inequality.

29 decreased by 26

A) $29 - 26$ B) 29^3

C) $26 - 29$ D) $\dfrac{29}{26}$

2) Write each as a verbal expression or equation or inequality.

12 (10)

A) 10 more than 12

B) the 12th power of 10

C) 12 less than 10

D) 10 times 12

3) Simplify the below expression.

$(8 + 2)^2 + 60 \div (10 - 4)$

A) 115 B) 99

C) 105 D) 110

4) Express the below as algebraic expression or equation or inequality.

the product of 4 and q

A) $q - 4$ B) $q + 4$

C) $4q$ D) $\dfrac{q}{4}$

5) Which expression is equivalent to

$1 - 16x + 1 + 12x$

6) Find the value of v :

$v - (-27) = 2$

A) -54 B) 29

C) $-\dfrac{2}{27}$ D) -25

7) Express the below as algebraic expression or equation or inequality.

the quotient of 15 and n

A) $15 + n$ B) $\dfrac{15}{n}$

C) $\dfrac{n}{15}$ D) $15n$

8) Write each as a verbal expression or equation or inequality.

$x + 10 < 20$

A) x decreased by 10 is less than 20

B) x increased by 10 is less than 20

C) 10 squared is less than 20

D) half of x is less than 20

9) Evaluate the below equation.

$$|x - 7| = 2$$

A) $-8, 8$ B) $9, 5$

C) $6, -6$ D) $23, -5$

10) Simplify the below expression.

$$(13)(10 + 7) - (10 - 4)^2$$

A) 185   B) 176

C) 166   D) 170

11) Write each as a verbal expression or equation or inequality.

$$a + 7$$

A) the product of a and 7
B) 7 more than a
C) 7 minus a
D) twice a

12) Simplify the below expression.

$$(25 + (7)(2) - 1 + 42) \div 16$$

A) 22   B) 15

C) 11   D) 5

13) Write each as a verbal expression or equation or inequality.

$$z - 8 < 30$$

A) half of z is less than 30

B) the difference of z and 8 is less than 30

C) the 8th power of z is less than 30

D) twice 8 is less than 30

14) Which expression is equivalent to

$$1 + 4b + 1 - 8b$$

15) Which expression is equivalent to

$$18n + 2n$$

16) Simplify the below expression.

$$20 - 4 + (2 + 6 + 8) \div 16$$

A) 24   B) 12

C) 11   D) 17

www.math-knots.com | www.a4ace.com

17) Evaluate the below expression, with the values given

$$h \div 4 + k - \left(j^2 - k\right)$$

Where $h = 8$, $j = 5$, and $k = 15$

A) 18   B) 0

C) 7   D) 6

18) Evaluate the below expression, with the values given

$$m + 10 + n + (n - n) \div 6$$

Where $m = 13$, and $n = 19$

A) 25   B) 39

C) 42   D) 32

19) Which expression is equivalent to

$$-8(-3p + 1) + p$$

20) Evaluate the below equation.

$$\left|-9 - b\right| = 13$$

A) $-11, 5$ B) $-22, 4$

C) -22 D) $5, -9$

21) Which expression is equivalent to

$$-11b - 13b$$

22) Which expression is equivalent to

$$-7(5x - 4) - 6(x - 5)$$

23) Which expression is equivalent to

$$3(1 + 7p) + 5(12p + 10)$$

24) Find the value of v :

$$-19.9v = 788.04$$

A) -39.6 B) -2.7

C) -3.8 D) 36.8

25) Which expression is equivalent to

0.5n - 3.1n

26) Find the value of 'n' that satisfies the below equation.

$3(2n+1)-(7+6n)=-4$

A) All real numbers.

B) −3

C) 16

D) 13

27) Evaluate the below expression, with the values given

$(19+x)(16-(z+y))-z$

Where $x=17$, $y=3$, and $z=8$

A) 172   B) 181

C) 156   D) 190

28) Find the value of 'x' that satisfies the below equation.

$7(1+4x)+2(6+5x)=57$

A) 1

B) −10

C) No solution.

D) All real numbers.

29) Which expression is equivalent to

$$\frac{1}{3}n+\frac{7}{4}\left(n-\frac{2}{3}\right)$$

30) Find the value of 'x' that satisfies the below equation.

$-5.88=-7.7x+5.6x$

A) 2.8   B) −8

C) 14   D) −6.387

31) Solve the equation for the indicated variable.

$kx=vw+yx$, for x

A) $x=\dfrac{k+y}{vw}$ B) $x=\dfrac{vw}{k-y}$

C) $x=\dfrac{vw}{-k+y}$ D) $x=\dfrac{vw}{k+y}$

32) Which expression is equivalent to

$-4(1+4r)$

www.math-knots.com | www.a4ace.com

33) Which expression is equivalent to

$$-5(1.4x - 9.2)$$

34) Find the value of k :

$$\frac{k}{10} = -32$$

A) $-\frac{16}{5}$ B) -320

C) -22 D) -42

35) Which expression is equivalent to

$$3.8 - 2.7x - 2 - 3x$$

36) Evaluate the below expression, with the values given

$$\left(a - \left(a - \left(a - c^2\right)\right)\right) \div 3$$

Where $a = 10$, and $c = 1$

A) 2   B) 11

C) 3   D) 15

37) Solve the equation for the indicated variable.

$$x - m = n - p + yx \text{, for } x$$

A) $x = (m + n - p)(-y - 1)$

B) $x = \dfrac{-m + n + p}{y + 1}$

C) $x = \dfrac{m + n - p}{-y + 1}$

D) $x = (m + n - p)(y + 1)$

38) Which expression is equivalent to

$$3(3x + 6)$$

39) Find the value of x :

$$12x + 5(5 + 3x) = -5(1 - 9x) - 3x$$

A) -8 B) 6

C) 2   D) 18

www.math-knots.com | www.a4ace.com

40) Which expression is equivalent to

$$5(1-2n)+12n$$

41) Find the value of x :

$$126 = -7x$$

A) −18   B) −28

C) −6   D) 6

42) Which expression is equivalent to

$$-2.6(4.644+7.5x)$$

43) Find the value of n :

$$21n = -399$$

A) −19   B) 17

C) 20   D) −26

44) Which expression is equivalent to

$$10(5x+9)-12$$

45) Express the below as algebraic expression or equation or inequality.

the product of b and 6

A) $6-b$ B) $b \cdot 6$
C) b^3 D) $6-b \geq 39$

46) Evaluate the below equation.

$$\frac{|3a|}{5} = 4$$

A) $\frac{20}{3}, -\frac{20}{3}$ B) 6, −4

C) 6, −6   D) 10, −10

47) Find the value of 'a' that satisfies the below equation.

$$5.57 = a + 1.77 - 3a$$

A) All real numbers.

B) 4.4

C) −13.564

D) −1.9

48) Find the value of n :

$-29.4 = n + (- 19)$

A) 27.5   B) −12.5

C) −10.4   D) −14.3

49) Find the value of z :

$- 11.9 = z - (- 19.9)$

A) −27.1   B) −31.8

C) 8   D) 28.4

50) Find the value of 'x' that satisfies the below equation.

$$-\frac{5}{6}x + \frac{43}{5}x = \frac{233}{50}$$

A) $\frac{3}{5}$ B) $\frac{13}{14}$

C) $-1\frac{7}{12}$ D) 10

51) Find the value of p :

$7 p - 12 (p + 4) = - 12 (p - 10)$

A) 24   B) 7

C) 17   D) − 16

52) Find the value of p :

$- 113 = - 5 + 3 (8 p - 4)$

53) Find the value of x :

$5 x + 6 (2 x + 4) = 143$

54) Find the value of m :

$- 2 (5 - 8 m) = - 106$

55) Evaluate the below equation.

$$|x - 10| - 1 = 7$$

A) −9   B) 18, 2

C) 18   D) 9 , −9

56) Find the value of k :

$$2(-8-6k)+12k=-4(k+6)+4$$

A) −12   B) −1

C) 10   D) −7

57) Find the value of n :

$$-5(1-4n)-12=5(4n+4)-7$$

A) 14   B) No solution.

C) 11   D) −20

58) Find the value of a:

$$3(5a-15)=-90+10a$$

A) −9

B) −29

C) −30

D) All real numbers.

59) Which expression is equivalent to

5n - 0.2n

60) Find the value of 'm' that satisfies the below equation.

$$m+1.3-7.4=-0.5$$

A) All real numbers.

B) 5.6

C) −4.7

D) −1.3

61) Evaluate the below equation.

$$\left|-3p\right|+2=20$$

A) 5, −7   B) −6, 6

C) 2, −14   D) 2

62) Find the value of x :

$$-18.5=8.2+x$$

A) −8.5   B) −9.83

C) −26.7   D) −38.2

63) Solve the equation for the indicated variable.

$$c-x=yx-dr, \text{ for } x$$

A) $x=\dfrac{c+dr}{y+1}$ B) $x=\dfrac{c-dr}{y-1}$

C) $x=\dfrac{y+1}{c+dr}$ D) $x=\dfrac{-c+dr}{-y+1}$

64) Find the value of 'b' that satisfies the below equation.

$$10 = 8(4b - 8) - 6(1 - 8b)$$

A) 1   B) −4

C) 11   D) 9

65) Which expression is equivalent to

$$m - 1.5 + 3.6$$

66) Find the value of 'n' that satisfies the below equation.

$$9 = 3(4n - 4) + 3(-4 + 7n)$$

A) −5   B) 1

C) 0   D) −16

67) Find the value of 'n' that satisfies the below equation.

$$-\frac{109}{56} = \frac{3}{2}n - \frac{7}{4} + \frac{7}{8}$$

A) $6\frac{3}{8}$ B) $-\frac{5}{7}$

C) $-\frac{11}{9}$ D) $-\frac{18}{13}$

68) Evaluate the below equation.

$$\left|\frac{n}{5}\right| = 1$$

A) 0, −8   B) 1, −7

C) 5, −5   D) 1

69) Evaluate the below equation.

$$|3 + v| = 11$$

A) 1   B) 8

C) 8, −14   D) 1, −7

70) Evaluate the below equation.

$$|6v| = 54$$

A) 6   B) 14, −14

C) 6, 0   D) 9, −9

71) Evaluate the below equation.

$$|-5x| = 50$$

A) −10, 10   B) 16, −10

C) 15, −15   D) 16

72) Evaluate the below equation.

$$|-7x + 1| = 50$$

A) 6, −18

B) $-\dfrac{28}{3}$, 10

C) 6, $-\dfrac{19}{2}$

D) $-7, \dfrac{51}{7}$

73) Write each as a verbal expression or equation or inequality.

$$n^3 = 9$$

A) 3 plus n is equal to 9

B) 3 more than n is equal to 9

C) n cubed is equal to 9

D) 3 cubed is equal to 9

74) Evaluate the below equation.

$$|-5p - 7| = 38$$

A) 7

B) $-9, \dfrac{31}{5}$

C) 1, 5

D) 7, −9

75) Which expression is equivalent to

$$- 9 x - 6 (1 + 11 x)$$

76) Evaluate the below equation.

$$|7 + 2b| = 3$$

A) −2

B) −2 , 6

C) −2 , −5

D) $9, -\dfrac{95}{9}$

77) Which expression is equivalent to

$$-\frac{1}{4}p + \frac{4}{5}\left(p - \frac{5}{2}\right)$$

78) Find the value of 'p' that satisfies the below equation.

$$0.5 = 1 - 4.7\,p + 4.5\,p$$

A) 2.5

B) −4.2

C) No solution.

D) − 14.2

79) Which expression is equivalent to

$$- 3 (1 - 10 x)$$

80) Evaluate the below equation.

$$3\left|\frac{x}{3}\right| = 3$$

A) 11, 1

B) 3, -3

C) 31, -29

D) -7, 7

81) Evaluate the below equation.

$$9 + \left|-9x\right| = 27$$

A) -2, 2

B) -9

C) -9, 9

D) 6, -6

82) Solve the equation for the indicated variable.

$$z = a\,m + a\,n + a\,p \text{ , for } a$$

A) $a = \dfrac{z}{m + n + p}$

B) $a = \dfrac{-m - n - p}{z}$

C) $a = -\dfrac{z}{-m - n + p}$

D) $a = z + m + n - p$

 **Algebra 1**

1) For which of the following values of n is this inequality is true ?

$$-33n > 99$$

A) $n < -3$ B) $n > -3$

C) $n > 132$ D) $n < 132$

2) For which of the following values of k is this inequality is true ?

$$-24 \leq 4k$$

A) $k \geq -6$ B) $k \geq 6$

C) $k \geq -28$ D) $k \geq -96$

3) For which of the following values of a is this inequality is true ?

$$11a \leq 385$$

A) $a \leq 35$ B) $a \leq 4235$

C) $a \geq 374$ D) $a \leq 374$

4) For which of the following values of v is this inequality is true ?

$$29 < \frac{v}{8}$$

A) $v > 232$ B) $v > -\frac{29}{8}$

C) $v < 37$ D) $v < -\frac{29}{8}$

5) For which of the following values of n is this inequality is true ?

$$2100 < 42n$$

A) $n > -50$ B) $n > 50$

C) $n > 88200$ D) $n < 88200$

6) For which of the following values of x is this inequality is true ?

$$22 \geq x + 44$$

A) $x \leq \frac{1}{2}$ B) $x \geq \frac{1}{2}$

C) $x \leq -22$ D) $x \leq -66$

7) For which of the following values of n is this inequality is true ?

$$\frac{n}{41} < 38$$

A) $n < 3$ B) $n > \frac{38}{41}$

C) $n < 1558$ D) $n < \frac{38}{41}$

8) For which of the following values of n is this inequality is true ?

$$63 \geq n + 33$$

A) $n \geq 30$ B) $n \leq \frac{21}{11}$

C) $n \leq 30$ D) $n \geq \frac{21}{11}$

9) For which of the following values of n is this inequality is true ?

$$\frac{n}{4} \geq -13$$

A) $n \geq 17$ B) $n \geq -52$

C) $n \leq 17$ D) $n \leq 9$

10) For which of the following values of b is this inequality is true ?

$$12b > -372$$

A) $b > -360$ B) $b < -360$

C) $b > -31$ D) $b > 360$

Algebra 1

11) For which of the following values of n is this inequality is true ?

$$20 > 16 - n$$

A) $n > -4$ B) $n > \dfrac{5}{4}$

C) $n > 320$ D) $n > -36$

12) For which of the following values of b is this inequality is true ?

$$-28 \geq -40 + b$$

A) $b \leq 12$ B) $b \geq 1120$

C) $b \geq \dfrac{7}{10}$ D) $b \geq 12$

13) For which of the following values of n is this inequality is true ?

$$n - 50 \geq -16$$

A) $n \leq -800$ B) $n \leq 66$

C) $n \geq 66$ D) $n \geq 34$

14) For which of the following values of x is this inequality is true ?

$$-22 \geq -11 - x$$

A) $x \geq -11$ B) $x \geq 242$

C) $x \geq 2$ D) $x \geq 11$

15) For which of the following values of n is this inequality is true ?

$$-41 + n > -67$$

A) $n > -26$ B) $n < -26$

C) $n > -108$ D) $n > \dfrac{67}{41}$

16) For which of the following values of n is this inequality is true ?

$$-1470 \leq -30n$$

A) $n \leq 49$ B) $n \leq -1500$

C) $n \leq -1440$ D) $n \leq 1440$

17) For which of the following values of r is this inequality is true ?

$$-14 + r > 31$$

A) $r < -\dfrac{31}{14}$ B) $r > 45$

C) $r < 17$ D) $r < 45$

18) For which of the following values of x is this inequality is true ?

$$29 \leq \dfrac{x}{27}$$

A) $x \leq 2$ B) $x \geq 783$

C) $x \geq 2$ D) $x \geq -\dfrac{29}{27}$

19) For which of the following values of x is this inequality is true ?

$$-16 \geq x - 17$$

A) $x \leq -33$ B) $x \leq 1$

C) $x \geq -33$ D) $x \geq 1$

20) For which of the following values of p is this inequality is true ?

$$-19 > p - 48$$

A) $p > \dfrac{19}{48}$ B) $p < \dfrac{19}{48}$

C) $p < -67$ D) $p < 29$

21) For which of the following values of x is this inequality is true ?

$$-2x \leq 42$$

A) $x \geq -21$ B) $x \leq 44$

C) $x \leq 40$ D) $x \leq -21$

22) For which of the following values of r is this inequality is true ?

$$-44 \leq \frac{r}{24}$$

A) $r \geq -\frac{11}{6}$ B) $r \geq -1056$

C) $r \geq 20$ D) $r \leq 20$

23) For which of the following values of m is this inequality is true ?

$$-4 \leq m - 31$$

A) $m \geq -\frac{4}{31}$ B) $m \leq -27$

C) $m \geq 27$ D) $m \leq -\frac{4}{31}$

24) For which of the following values of m is this inequality is true ?

$$m - 14 \geq 4$$

A) $m \leq 18$ B) $m \geq \frac{2}{7}$

C) $m \geq 18$ D) $m \leq \frac{2}{7}$

25) For which of the following values of n is this inequality is true ?

$$5 > \frac{n}{31}$$

A) $n < \frac{5}{31}$ B) $n < 36$

C) $n < -26$ D) $n < 155$

26) For which of the following values of x is this inequality is true ?

$$12 < 12 + x$$

A) $x < 144$ B) $x > 144$

C) $x < 0$ D) $x > 0$

27) For which of the following values of v is this inequality is true ?

$$70 < 20 + v$$

A) $v > 50$ B) $v > \frac{7}{2}$

C) $v > -90$ D) $v > 1400$

28) For which of the following values of x is this inequality is true ?

$$2x < 66$$

A) $x < 132$ B) $x < 33$

C) $x > 68$ D) $x < 68$

29) For which of the following values of x is this inequality is true ?

$$31 < \frac{x}{50}$$

A) $x > 1550$ B) $x < 1550$

C) $x < 81$ D) $x < -19$

30) For which of the following values of a is this inequality is true ?

$$-35 < \frac{a}{9}$$

A) $a > \frac{35}{9}$ B) $a > -26$

C) $a > -315$ D) $a > -\frac{35}{9}$

www.math-knots.com | www.a4ace.com

31) For which of the following values of x is this inequality is true ?

$$-35x \leq 630$$

A) $x \geq 595$ B) $x \geq 18$

C) $x \geq -22050$ D) $x \geq -18$

32) For which of the following values of k is this inequality is true ?

$$\frac{k}{21} < 47$$

A) $k < -26$ B) $k < -987$

C) $k < 26$ D) $k < 987$

33) For which of the following values of n is this inequality is true ?

$$-459 > -27n$$

A) $n < -432$ B) $n > 17$

C) $n < 17$ D) $n > -432$

34) For which of the following values of x is this inequality is true ?

$$29x > 812$$

A) $x > -23548$ B) $x > 23548$

C) $x > 841$ D) $x > 28$

35) For which of the following values of n is this inequality is true ?

$$36 + n \leq 14$$

A) $n \leq \frac{7}{18}$ B) $n \leq -22$

C) $n \leq 50$ D) $n \geq \frac{7}{18}$

36) For which of the following values of a is this inequality is true ?

$$47a \leq -846$$

A) $a \leq 799$ B) $a \geq 799$

C) $a \geq -893$ D) $a \leq -18$

37) For which of the following values of p is this inequality is true ?

$$-31p \geq 1271$$

A) $p \leq -39401$ B) $p \leq -41$

C) $p \geq -41$ D) $p \geq -39401$

38) For which of the following values of n is this inequality is true ?

$$n + 46 \geq 76$$

A) $n \leq 122$ B) $n \leq \frac{38}{23}$

C) $n \geq 30$ D) $n \leq 30$

39) For which of the following values of b is this inequality is true ?

$$0 \leq \frac{b}{31}$$

A) $b \leq 0$ B) $b \leq 31$

C) $b \geq 31$ D) $b \geq 0$

40) For which of the following values of b is this inequality is true ?

$$\frac{b}{9} < 16$$

A) $b < 144$ B) $b > -7$

C) $b > \frac{16}{9}$ D) $b > 144$

41) For which of the following values of x is this inequality is true ?

$$35x \le 1645$$

A) $x \le 57575$ B) $x \le -1680$

C) $x \le 47$ D) $x \le 1610$

42) For which of the following values of n is this inequality is true ?

$$n - 5 \le 39$$

A) $n \ge 34$ B) $n \le 195$

C) $n \le 44$ D) $n \le 34$

43) For which of the following values of n is this inequality is true ?

$$-17 < -27n$$

A) $n > -\dfrac{17}{16}$ B) $n < -\dfrac{17}{27}$

C) $n > -1$ D) $n > 272$

44) For which of the following values of x is this inequality is true ?

$$27 \le 1 - x$$

A) $x \le -26$ B) $x \le 26$

C) $x \le -27$ D) $x \le 27$

45) For which of the following values of n is this inequality is true ?

$$x - 22 \ge -10$$

A) $x \le -32$ B) $x \ge 12$

C) $x \le 12$ D) $x \le -220$

46) For which of the following values of r is this inequality is true ?

$$-465 > -31r$$

A) $r < 496$ B) $r > 14415$

C) $r > 496$ D) $r > 15$

47) For which of the following values of n is this inequality is true ?

$$-14 < 14n$$

A) $n > 0$ B) $n > 196$

C) $n > -1$ D) $n > -28$

48) For which of the following values of k is this inequality is true ?

$$k + 12 < 29$$

A) $k < 17$ B) $k > 17$

C) $k > \dfrac{29}{12}$ D) $k > 41$

49) For which of the following values of n is this inequality is true ?

$$-928 < 32n$$

A) $n > -29$ B) $n > 29696$

C) $n > -960$ D) $n < -29$

50) For which of the following values of n is this inequality is true ?

$$n + 16 > -24$$

A) $n < 384$ B) $n > 384$

C) $n > -40$ D) $n > -384$

51) For which of the following values of x is this inequality is true ?

$$6x < 186$$

A) x < 31   B) x > − 31

C) x > 31   D) x < − 31

52) For which of the following values of a is this inequality is true ?

$$a - 37 < - 50$$

A) a > − 13   B) a < − 87

C) a < − 13   D) a < − 1850

53) For which of the following values of x is this inequality is true ?

$$\frac{x}{31} < 3$$

A) x < 34   B) $x > \frac{3}{31}$

C) x < 93   D) $x < \frac{3}{31}$

54) For which of the following values of n is this inequality is true ?

$$n - 6 \geq -11$$

A) n ≥ −5   B) n ≥ −17

C) $n \geq \frac{11}{6}$ D) n ≥ 66

55) For which of the following values of v is this inequality is true ?

$$43v \leq 1677$$

A) v ≤ 72111   B) v ≥ −72111

C) v ≥ 72111   D) v ≤ 39

56) For which of the following values of k is this inequality is true ?

$$k + 38 > 82$$

A) k > 44   B) k < 44

C) k > 120   D) k < 120

57) For which of the following values of x is this inequality is true ?

$$989 > - 43x$$

A) x > − 1032   B) x > 946

C) x > 1032   D) x > − 23

58) For which of the following values of r is this inequality is true ?

$$\frac{r}{7} < 49$$

A) r < − 42   B) r > − 42

C) r < 343   D) r > 7

59) For which of the following values of n is this inequality is true ?

$$n - (-9) \geq -11$$

A) n ≥ 2   B) n ≥ −20

C) n ≤ 2   D) n ≥ −99

60) For which of the following values of a is this inequality is true ?

$$154 \leq -11a$$

A) a ≥ −1694   B) a ≥ −14

C) a ≥ 143   D) a ≤ −14

61) For which of the following values of x is this inequality is true ?

$$26x > 728$$

A) $x < 754$ B) $x > 28$

C) $x < 28$ D) $x > 754$

62) For which of the following values of n is this inequality is true ?

$$3 \ge n - (-32)$$

A) $n \le -96$ B) $n \le -29$

C) $n \ge -35$ D) $n \le -35$

63) For which of the following values of k is this inequality is true ?

$$k + 9 \le 56$$

A) $k \le 65$ B) $k \ge 65$

C) $k \le \dfrac{56}{9}$ D) $k \le 47$

64) For which of the following values of x is this inequality is true ?

$$2 \le x - 10$$

A) $x \le 12$ B) $x \le -8$

C) $x \ge 12$ D) $x \ge -8$

65) For which of the following values of x is this inequality is true ?

$$\dfrac{x}{46} \le -15$$

A) $x \le -61$ B) $x \le -\dfrac{15}{46}$

C) $x \ge -690$ D) $x \le -690$

66) For which of the following values of b is this inequality is true ?

$$29 - b \ge 59$$

A) $b \le 30$ B) $b \ge -88$

C) $b \le -88$ D) $b \le -30$

67) For which of the following values of x is this inequality is true ?

$$-2 > \dfrac{x}{49}$$

A) $x < -98$ B) $x < 51$

C) $x < -\dfrac{2}{49}$ D) $x < \dfrac{2}{49}$

68) For which of the following values of n is this inequality is true ?

$$\dfrac{23}{43} > \dfrac{n}{43}$$

A) $n < -\dfrac{1826}{43}$ B) $n > -\dfrac{23}{1849}$

C) $n > -\dfrac{1826}{43}$ D) $n < 23$

69) For which of the following values of k is this inequality is true ?

$$k - (-18) > -2$$

A) $k < -\dfrac{1}{9}$ B) $k > 36$

C) $k > -20$ D) $k > -\dfrac{1}{9}$

70) For which of the following values of x is this inequality is true ?

$$1254 < 38x$$

A) $x > 33$ B) $x > -47652$

C) $x < 33$ D) $x > -1216$

71) For which of the following values of a is this inequality is true ?

$$\frac{a}{17} < 19$$

A) $a < \frac{19}{17}$ B) $a < 323$

C) $a < 2$ D) $a > \frac{19}{17}$

72) For which of the following values of x is this inequality is true ?

$$36x > -936$$

A) $x > -26$ B) $x < -26$

C) $x > -900$ D) $x > 972$

73) For which of the following values of k is this inequality is true ?

$$k + 1 < -6$$

A) $k > -7$ B) $k > -5$

C) $k < -7$ D) $k < -6$

74) For which of the following values of m is this inequality is true ?

$$-2m < -12$$

A) $m < 6$ B) $m < 24$

C) $m > 6$ D) $m < 10$

75) For which of the following values of x is this inequality is true ?

$$22 \geq x + (-23)$$

A) $x \leq 506$ B) $x \leq -1$

C) $x \geq -1$ D) $x \leq 45$

76) For which of the following values of m is this inequality is true ?

$$-144 \leq -6m$$

A) $m \leq 864$ B) $m \leq -24$

C) $m \geq 24$ D) $m \leq 24$

77) For which of the following values of x is this inequality is true ?

$$\frac{x}{14} < 47$$

A) $x < 33$ B) $x < \frac{47}{14}$

C) $x < 658$ D) $x > \frac{47}{14}$

78) For which of the following values of k is this inequality is true ?

$$-13 \geq k - 42$$

A) $k \geq -29$ B) $k \geq -55$

C) $k \geq 29$ D) $k \leq 29$

79) For which of the following values of k is this inequality is true ?

$$-23 \geq x - (-22)$$

A) $x \leq -45$ B) $x \geq 506$

C) $x \geq -1$ D) $x \geq -45$

80) For which of the following values of x is this inequality is true ?

$$r - (-50) \leq 51$$

A) $r \leq -1$ B) $r \leq 1$

C) $r \leq \frac{51}{50}$ D) $r \leq 101$

81) For which of the following values of b is this inequality is true ?

$$-33 < b + (-17)$$

A) b > −50   B) b > 50

C) b < −50   D) b > −16

82) For which of the following values of r is this inequality is true ?

$$\frac{10}{29} \le \frac{r}{29}$$

A) $r \le \frac{10}{841}$ B) $r \ge 10$

C) $r \ge \frac{10}{841}$ D) $r \ge \frac{851}{29}$

83) For which of the following values of x is this inequality is true ?

$$1056 \le 24x$$

A) x ≥ 25344   B) x ≥ 44

C) x ≤ −44   D) x ≤ 25344

84) For which of the following values of x is this inequality is true ?

$$1296 < 36x$$

A) x > 1260   B) x > 36

C) x > 1332   D) x < 36

85) For which of the following values of x is this inequality is true ?

$$\frac{k}{23} \ge -47$$

A) k ≤ −1081   B) k ≤ 1081

C) k ≥ −1081   D) k ≤ −70

86) For which of the following values of p is this inequality is true ?

$$p - 44 > -73$$

A) p > −29   B) p > −3212

C) p > 29   D) p > −117

87) For which of the following values of r is this inequality is true ?

$$-513 \ge 19r$$

A) r ≤ −494   B) r ≥ −494

C) r ≥ −9747   D) r ≤ −27

88) For which of the following values of n is this inequality is true ?

$$26n > 650$$

A) n > 25   B) n > 16900

C) n > 676   D) n < 676

89) For which of the following values of x is this inequality is true ?

$$-8 > 33 + x$$

A) x < −41   B) x > −264

C) $x > -\frac{8}{33}$ D) $x < -\frac{8}{33}$

90) For which of the following values of m is this inequality is true ?

$$44m > 2156$$

A) m > 2200   B) m > 49

C) m > −2112   D) m < −2112

91) For which of the following values of a is this inequality is true ?

$$-20a \le -180$$

A) a ≤ 3600   B) a ≤ 9

C) a ≤ -160   D) a ≥ 9

92) For which of the following values of k is this inequality is true ?

$$-32 - k \ge -39$$

A) k ≤ 7   B) k ≤ -1248

C) k ≤ 1248   D) $k \le -\dfrac{39}{32}$

www.math-knots.com | www.a4ace.com

1) For which of the following values of n is this inequality and the graph is true ?

$$25.4 < 15 - n$$

A) $n < -40.4$:

B) $n < -10.4$:

C) $n < 10.4$:

D) $n > -10.4$:

4) For which of the following values of m is this inequality and the graph is true ?

$$23.9 > m + 12.5$$

A) $m < 11.4$:

B) $m < -11.4$:

C) $m > -11.4$:

D) $m < -36.4$:

2) For which of the following values of b is this inequality and the graph is true ?

$$-8.7 \geq -7.7 + b$$

A) $b \geq -1$:

B) $b \leq -1$:

C) $b \geq -66.99$:

D) $b \geq -1$:

5) For which of the following values of b is this inequality and the graph is true ?

$$-4.6b \geq 53.36$$

A) $b \geq 11.6$:

B) $b \leq -11.6$:

C) $b \leq 11.6$:

D) $b \geq -11.6$:

3) For which of the following values of n is this inequality and the graph is true ?

$$\frac{n}{8.1} > 18.7$$

A) $n < 26.8$:

B) $n < 151.47$:

C) $n < 26.8$:

D) $n > 151.47$:

6) For which of the following values of r is this inequality and the graph is true ?

$$r - (-15.3) < 30.3$$

A) $r < 15$:

B) $r > 15$:

C) $r > -15$:

D) $r < -15$:

7) For which of the following values of p is this inequality and the graph is true ?

$$\frac{p}{12.1} > -16.4$$

A) p > −198.44:

B) p > −198.44:

C) p > −28.5:

D) p > 4.3:

8) For which of the following values of p is this inequality and the graph is true ?

$$1.7 + p > 16.39$$

A) p > 14.69:

B) p > −14.69:

C) p < −18.09:

D) p < 14.69:

9) For which of the following values of x is this inequality and the graph is true ?

$$\frac{x}{8.8} \geq 17.8$$

A) x ≥ 156.64:

B) x ≤ 156.64:

C) x ≥ 26.6:

D) x ≤ 26.6:

10) For which of the following values of b is this inequality and the graph is true ?

$$\frac{b}{7.5} \leq 0.3$$

A) b ≤ −7.8:

B) b ≤ 2.25:

C) b ≤ 7.8:

D) b ≤ 7.8:

11) For which of the following values of n is this inequality and the graph is true ?

$$9.4 > 18.7 + n$$

A) n > 175.78:

B) n < −9.3:

C) n > −9.3:

D) n > 28.1:

12) For which of the following values of x is this inequality and the graph is true ?

$$\frac{x}{2.4} < -18.7$$

A) x < −44.88:

B) x > −21.1:

C) x > −44.88:

D) x < −16.3:

13) For which of the following values of x is this inequality and the graph is true ?

$$16.3 \, x < 52.16$$

A) x < −68.46:

   −72 −70 −68 −66 −64

B) x > 3.2:

   −4 −2 0 2 4 6

C) x < 3.2:

   −4 −2 0 2 4 6

D) x < 3.2:

   −4 −2 0 2 4 6

14) For which of the following values of k is this inequality and the graph is true ?

$$-1.7 \, k < 21.93$$

A) k > −805.089:

   −808 −804

B) k < −12.9:

   −14 −12 −10 −8 −6

C) k > −12.9:

   −14 −12 −10 −8 −6

D) k < −805.089:

   −808 −804

15) For which of the following values of n is this inequality and the graph is true ?

$$-8.5 + n \le -26.1$$

A) n ≥ 221.85:

   214 216 218 220 222

B) n ≤ 221.85:

   216 218 220 222 224 226

C) n ≥ 221.85:

   216 218 220 222 224 226

D) n ≤ −17.6:

   −22 −20 −18 −16 −14 −12

16) For which of the following values of p is this inequality and the graph is true ?

$$- 10.5 > 3.1 + p$$

A) p > −32.55:

   −40 −38 −36 −34 −32 −30

B) p < −13.6:

   −14 −12 −10 −8 −6

C) p > −13.6:

   −14 −12 −10 −8 −6

D) p > 32.55:

   28 30 32 34 36 38

17) For which of the following values of n is this inequality and the graph is true ?

$$\frac{n}{16.8} > 14.6$$

A) n > 245.28:

   242 244 246 248 250

B) n < 245.28:

   242 244 246 248 250

C) n < 2.2:

   −4 −2 0 2 4 6

D) n < 2.2:

   −4 −2 0 2 4 6

18) For which of the following values of x is this inequality and the graph is true ?

$$12.4 < \frac{x}{1.3}$$

A) x > 16.12:

   14 16 18 20 22

B) x < 13.7:

   6 8 10 12 14 16

C) x > 16.12:

   12 14 16 18 20

D) x < 16.12:

   12 14 16 18 20

Algebra 1

19) For which of the following values of x is this inequality and the graph is true ?

$$x + 17.4 \leq 27.6$$

A) $x \geq 480.24$:
476 478 480 482 484 486

B) $x \geq 10.2$:
2 4 6 8 10 12

C) $x \leq 10.2$:
2 4 6 8 10 12

D) $x \geq 10.2$:
2 4 6 8 10 12

20) For which of the following values of p is this inequality and the graph is true ?

$$10.5 + p > -6.9$$

A) $p < -72.45$
-78 -76 -74 -72 -70 -68

B) $p > -72.45$
-78 -76 -74 -72 -70 -68

C) $p > 72.45$:
70 72 74 76 78

D) $p > -17.4$:
-20 -18 -16 -14 -12

21) For which of the following values of n is this inequality and the graph is true ?

$$2 > 10.7 + n$$

A) $n > -8.7$:
-12 -10 -8 -6 -4

B) $n < -8.7$:
-12 -10 -8 -6 -4

C) $n > -8.7$:
-12 -10 -8 -6 -4

D) $n < 12.7$:
8 10 12 14 16

22) For which of the following values of a is this inequality and the graph is true ?

$$16 \leq a + 8.3$$

A) $a \leq 7.7$:
0 2 4 6 8

B) $a \geq 7.7$:
0 2 4 6 8

C) $a \geq 132.8$:
128 130 132 134 136

D) $a \geq -7.7$:
-10 -8 -6 -4 -2

23) For which of the following values of b is this inequality and the graph is true ?

$$0.77 > -11.73 - b$$

A) $b > -12.5$:
-18 -16 -14 -12 -10 -8

B) $b > -10.96$:
-14 -12 -10 -8 -6 -4

C) $b > 12.5$:
6 8 10 12 14

D) $b < -10.96$:
-16 -14 -12 -10 -8 -6

24) For which of the following values of b is this inequality and the graph is true ?

$$1.4 b > -22.12$$

A) $b > -20.72$:
-26 -24 -22 -20 -18 -16

B) $b > -20.72$:
-26 -24 -22 -20 -18

C) $b < -15.8$:
-22 -20 -18 -16 -14 -12

D) $b > -15.8$:
-22 -20 -18 -16 -14 -12

25) For which of the following values of x is this inequality and the graph is true ?

$$18.6 - x > 6.1$$

A) x > −24.7 :
$$\text{number line: } -30 \; -28 \; -26 \; -24 \; -22$$

B) x < 12.5 :
$$\text{number line: } 10 \; 12 \; 14 \; 16 \; 18$$

C) x > −12.5 :
$$\text{number line: } -16 \; -14 \; -12 \; -10 \; -8$$

D) x > 12.5 :
$$\text{number line: } 10 \; 12 \; 14 \; 16 \; 18$$

26) For which of the following values of r is this inequality and the graph is true ?

$$-33 < r + (-15.7)$$

A) r > 518.1 :
$$\text{number line: } 512 \; 514 \; 516 \; 518 \; 520 \; 522$$

B) r > −17.3 :
$$\text{number line: } -20 \; -18 \; -16 \; -14 \; -12$$

C) r > 518.1 :
$$\text{number line: } 512 \; 514 \; 516 \; 518 \; 520 \; 522$$

D) r > −48.7 :
$$\text{number line: } -54 \; -52 \; -50 \; -48 \; -46$$

27) For which of the following values of a is this inequality and the graph is true ?

$$-0.5a \le -3.8$$

A) a ≤ −3.3 :
$$\text{number line: } -8 \; -6 \; -4 \; -2 \; 0 \; 2$$

B) a ≤ 4.3 :
$$\text{number line: } 0 \; 2 \; 4 \; 6 \; 8$$

C) a ≥ −3.3 :
$$\text{number line: } -8 \; -6 \; -4 \; -2 \; 0 \; 2$$

D) a ≥ 7.6 :
$$\text{number line: } 2 \; 4 \; 6 \; 8 \; 10 \; 12$$

28) For which of the following values of x is this inequality and the graph is true ?

$$-18.036 \ge x - 6.436$$

A) x ≥ −11.6 :
$$\text{number line: } -16 \; -14 \; -12 \; -10 \; -8$$

B) x ≥ −11.6 :
$$\text{number line: } -18 \; -16 \; -14 \; -12 \; -10$$

C) x ≤ 7 :
$$\text{number line: } 4 \; 6 \; 8 \; 10 \; 12 \; 14$$

D) x ≤ −11.6 :
$$\text{number line: } -18 \; -16 \; -14 \; -12 \; -10$$

29) For which of the following values of n is this inequality and the graph is true ?

$$-1.81 \ge -6.41 + n$$

A) n ≥ −8.22 :
$$\text{number line: } -14 \; -12 \; -10 \; -8 \; -6 \; -4$$

B) n ≥ 4.6 :
$$\text{number line: } 2 \; 4 \; 6 \; 8 \; 10$$

C) n ≥ 4.6 :
$$\text{number line: } -4 \; -2 \; 0 \; 2 \; 4 \; 6$$

D) n ≤ 4.6 :
$$\text{number line: } 2 \; 4 \; 6 \; 8 \; 10$$

30) For which of the following values of x is this inequality and the graph is true ?

$$\frac{x}{9.9} \ge 8.8$$

A) x ≤ 18.7 :
$$\text{number line: } 16 \; 18 \; 20 \; 22 \; 24$$

B) x ≥ 87.12 :
$$\text{number line: } 80 \; 82 \; 84 \; 86 \; 88$$

C) x ≤ 87.12 :
$$\text{number line: } 80 \; 82 \; 84 \; 86 \; 88$$

D) x ≥ 18.7 :
$$\text{number line: } 16 \; 18 \; 20 \; 22 \; 24$$

 www.math-knots.com | www.a4ace.com

31) For which of the following values of v is this inequality and the graph is true ?

$$v + (-1.53) \le 6.47$$

A) $v \le 8$: ![number line with closed circle at 8, shaded left; marks 4 6 8 10 12 14]

B) $v \le 4.94$: ![number line with open circle at 5, shaded left; marks -2 0 2 4 6 8]

C) $v \le 8$: ![number line with open circle at 8, shaded right; marks 4 6 8 10 12 14]

D) $v \le 8$: ![number line with open circle at 8, shaded right; marks 4 6 8 10 12 14]

32) For which of the following values of p is this inequality and the graph is true ?

$$-41 \le -5p$$

A) $p \ge -36$: ![number line closed circle at -36 shaded right; marks -38 -36 -34 -32 -30 -28]

B) $p \le -36$: ![number line open circle at -36 shaded right; marks -38 -36 -34 -32 -30 -28]

C) $p \le 8.2$: ![number line closed circle at 8 shaded left; marks 2 4 6 8 10 12]

D) $p \ge -\dfrac{41}{5}$: ![number line closed circle at -8 shaded right; marks -12 -10 -8 -6 -4]

33) For which of the following values of a is this inequality and the graph is true ?

$$0.5a \ge 1.65$$

A) $a \ge 1.15$: ![number line open circle at 1 shaded right; marks -4 -2 0 2 4 6]

B) $a \ge 3.3$: ![number line closed circle at 3 shaded right; marks -4 -2 0 2 4 6]

C) $a \le 1.15$: ![number line closed circle at 1 shaded left; marks -4 -2 0 2 4 6]

D) $a \le 3.3$: ![number line open circle at 3 shaded left; marks -2 0 2 4 6]

34) For which of the following values of p is this inequality and the graph is true ?

$$-0.6 - p < 16.86$$

A) $p > -17.46$: ![number line open circle at -17 shaded right; marks -20 -18 -16 -14 -12 -10]

B) $p > 17.46$: ![number line closed circle at 17 shaded right; marks 14 16 18 20 22]

C) $p < 17.46$: ![number line open circle at 17 shaded right; marks 14 16 18 20 22]

D) $p > 17.46$: ![number line open circle at 17 shaded right; marks 14 16 18 20 22]

35) For which of the following values of x is this inequality and the graph is true ?

$$12.4 < 3 - x$$

A) $x > 15.4$: ![number line open circle at 15 shaded right; marks 10 12 14 16 18 20]

B) $x > 15.4$: ![number line closed circle at 15 shaded right; marks 10 12 14 16 18 20]

C) $x < -9.4$: ![number line open circle at -9 shaded left; marks -16 -14 -12 -10 -8 -6]

D) $x < 15.4$: ![number line open circle at 15 shaded left; marks 12 14 16 18 20]

36) For which of the following values of n is this inequality and the graph is true ?

$$\dfrac{n}{7.4} \le 16.7$$

A) $n \le -123.58$: ![number line closed circle at -123 shaded left; marks -128 -124 -120]

B) $n \le 9.3$: ![number line closed circle at 9 shaded left; marks 4 6 8 10 12 14]

C) $n \ge -123.58$: ![number line closed circle at -123 shaded right; marks -128 -124 -120]

D) $n \le 123.58$: ![number line closed circle at 123 shaded left; marks 118 120 122 124 126]

www.math-knots.com | www.a4ace.com

37) For which of the following values of p is this inequality and the graph is true ?

$$-10.9 > \frac{p}{11.1}$$

A) p < −120.99:

B) p > −120.99:

C) p > 120.99:

D) p > −0.2:

38) For which of the following values of p is this inequality and the graph is true ?

$$-8.5\,p < -70.55$$

A) p > 8.3:

B) p > 62.05:

C) p > 8.3:

D) p < 8.3:

39) For which of the following values of x is this inequality and the graph is true ?

$$\frac{x}{4.7} < 4.7$$

A) x < −1:

B) x < 22.09:

C) x < 9.4:

D) x < 0:

40) For which of the following values of x is this inequality and the graph is true ?

$$20.2 < 1.6 + x$$

A) x > 18.6:

B) x < 18.6:

C) x < 18.6:

D) x < 32.32:

41) For which of the following values of b is this inequality and the graph is true ?

$$-3.186 < b - 12.086$$

A) b > 8.9:

B) b < 8.9:

C) b < 8.9:

D) b < −8.9:

42) For which of the following values of v is this inequality and the graph is true ?

$$v + 18.8 > 12.9$$

A) v < 31.7:

B) v < −5.9:

C) v < −5.9:

D) v > −5.9:

43) For which of the following values of x is this inequality and the graph is true ?

$$\frac{x}{7.9} \le -5.6$$

A) x ≤ −44.24:

B) x ≤ −2.3:

C) x ≤ 13.5:

D) x ≥ 13.5:

44) For which of the following values of n is this inequality and the graph is true ?

$$75.68 > 4.4\,n$$

A) n > −71.28:

B) n > 17.2:

C) n > 17.2:

D) n < 17.2:

45) For which of the following values of m is this inequality and the graph is true ?

$$-13.6 \ge \frac{m}{7.7}$$

A) m ≤ −104.72:

B) m ≤ −21.3:

C) m ≥ −104.72:

D) m ≥ −21.3:

46) For which of the following values of n is this inequality and the graph is true ?

$$15.66 > -2.7\,n$$

A) n < −5.8:

B) n > −5.8:

C) n < 12.96:

D) n < 12.96:

47) For which of the following values of r is this inequality is true ?

$$\frac{r + 8}{-2} \ge -2$$

A) r ≤ −92   B) r ≤ 20

C) r ≥ 20   D) r ≤ −4

48) For which of the following values of p is this inequality is true ?

$$5 + \frac{p}{-7} \ge 0$$

A) p ≤ −52   B) p ≥ 35

C) p ≤ 35   D) p ≤ −27

www.math-knots.com | www.a4ace.com

49) For which of the following values of r is this inequality and the graph is true ?

$$450 \geq 15(r - 2)$$

A) $r \geq -90$ B) $r \leq 32$

C) $r \leq -90$ D) $r \geq 32$

50) For which of the following values of r is this inequality is true ?

$$-7 < \frac{-8 + r}{4}$$

A) $r < 18$ B) $r > -20$

C) $r < -20$ D) $r > 18$

51) For which of the following values of b is this inequality is true ?

$$\frac{20 + b}{3} < 11$$

A) $b < 13$ B) $b > 11$

C) $b > 13$ D) $b > -93$

52) For which of the following values of n is this inequality is true ?

$$-45 > -15(n - 4)$$

A) $n > 7$ B) $n > -16$

C) $n > -63$ D) $n < 7$

53) For which of the following values of x is this inequality is true ?

$$404 > 18x + 8$$

A) $x < -12$ B) $x < 22$

C) $x < -78$ D) $x < -70$

54) For which of the following values of m is this inequality is true ?

$$9 + \frac{m}{2} \geq 7$$

A) $m \leq -11$ B) $m \leq -14$

C) $m \geq -4$ D) $m \geq -11$

55) For which of the following values of x is this inequality is true ?

$$\frac{x + 5}{-3} \geq 5$$

A) $x \leq -20$ B) $x \geq -14$

C) $x \geq -20$ D) $x \leq -14$

56) For which of the following values of a is this inequality is true ?

$$3 \leq \frac{a - 15}{-14}$$

A) $a \leq -8$ B) $a \leq -57$

C) $a \leq -39$ D) $a \leq -27$

57) For which of the following values of x is this inequality is true ?

$$\frac{14 + x}{20} < -1$$

A) $x < -39$ B) $x < -34$

C) $x < -21$ D) $x < -45$

58) For which of the following values of n is this inequality is true ?

$$\frac{-3 + n}{-6} < -2$$

A) $n < -18$ B) $n > 6$

C) $n > -18$ D) $n > 15$

59) For which of the following values of k is this inequality is true ?

$$18 < \frac{k}{-14} + 19$$

A) k < 1 B) k < 14

C) k < − 51 D) k > 14

60) For which of the following values of x is this inequality is true ?

$$-20 + \frac{x}{21} < -21$$

A) x > − 34 B) x > 9

C) x > − 21 D) x < − 21

61) For which of the following values of v is this inequality is true ?

$$159 \le -9 + 6v$$

A) v ≥ 28 B) v ≤ −11

C) v ≤ 20 D) v ≤ 28

62) For which of the following values of x is this inequality is true ?

$$16(x + 5) \le -176$$

A) x ≤ −44 B) x ≤ −66

C) x ≤ −16 D) x ≤ −25

63) For which of the following values of n is this inequality is true ?

$$\frac{n}{2} - 16 \ge -10$$

A) n ≥ −33 B) n ≥ 12

C) n ≥ 5 D) n ≥ 3

64) For which of the following values of m is this inequality is true ?

$$-187 < 11(m - 13)$$

A) m > − 4 B) m < − 73

C) m < − 4 D) m < 17

65) For which of the following values of n is this inequality is true ?

$$\frac{n}{16} - 9 > -8$$

A) n > 16 B) n > 8

C) n > − 20 D) n < − 20

66) For which of the following values of n is this inequality is true ?

$$-14(1 + n) \le -434$$

A) n ≤ −69 B) n ≥ −100

C) n ≥ −69 D) n ≥ 30

67) For which of the following values of b is this inequality is true ?

$$8(b - 8) \le -64$$

A) b ≤ _97 B) b ≥ 0

C) b ≥ −97 D) b ≤ 0

68) For which of the following values of v is this inequality is true ?

$$3 < \frac{v}{2} + 5$$

A) v > − 4 B) v > − 5

C) v > − 6 D) v > − 100

 **Algebra 1**

69) For which of the following values of x is this inequality is true ?

$$-3 < \frac{19 + x}{4}$$

A) x > − 31   B) x > − 4

C) x < 19   D) x > 19

70) For which of the following values of x is this inequality is true ?

$$\frac{x}{26} - 5 \le -4$$

A) x ≤ 26   B) x ≤ −82

C) x ≤ −51   D) x ≤ −49

71) For which of the following values of k is this inequality is true ?

$$-8 \ge \frac{k - 12}{4}$$

A) k ≤ −33   B) k ≤ −20

C) k ≤ 12   D) k ≤ 20

72) For which of the following values of k is this inequality is true ?

$$-12 > -14 + \frac{k}{-2}$$

A) k > −4   B) k < −4

C) k < −100   D) k < −38

73) For which of the following values of a is this inequality is true ?

$$4 \ge \frac{20 + a}{-2}$$

A) a ≥ −18   B) a ≥ −28

C) a ≥ −11   D) a ≥ −14

74) For which of the following values of n is this inequality is true ?

$$387 \le 9(n + 7)$$

A) n ≤ 36   B) n ≥ 36

C) n ≥ −99   D) n ≤ − 99

75) For which of the following values of v is this inequality is true ?

$$14(20 + v) > 630$$

A) v > 25   B) v < − 79

C) v > − 79   D) v < − 100

76) For which of the following values of n is this inequality is true ?

$$-88 > -16 + 6n$$

A) n < − 66   B) n < − 12

C) n > 11   D) n < − 66

77) For which of the following values of k is this inequality is true ?

$$351 > 13(k - 9)$$

A) k > 36   B) k < 36

C) k > − 9   D) k > − 51

78) For which of the following values of a is this inequality is true ?

$$-14 > \frac{-13 + a}{2}$$

A) a < − 12   B) a < 9

C) a < − 15   D) a < − 59

79) For which of the following values of k is this inequality is true ?

$$19 - 17k \geq 699$$

A) $k \leq -11$ B) $k \geq -11$

C) $k \geq -40$ D) $k \leq -40$

80) For which of the following values of x is this inequality is true ?

$$\frac{-12 + x}{5} < -4$$

A) $x > -54$ B) $x < -8$

C) $x < -54$ D) $x > -8$

81) For which of the following values of x is this inequality is true ?

$$-344 \geq -16x - 8$$

A) $x \leq 21$ B) $x \geq 21$

C) $x \geq -67$ D) $x \leq -67$

82) For which of the following values of p is this inequality is true ?

$$16 + \frac{p}{-4} \leq 8$$

A) $p \leq -92$ B) $p \geq -92$

C) $p \geq 32$ D) $p \leq -56$

83) For which of the following values of x is this inequality is true ?

$$8 < \frac{x + 10}{3}$$

A) $x < -22$ B) $x > -37$

C) $x > 14$ D) $x < -37$

84) For which of the following values of x is this inequality is true ?

$$-8 \leq \frac{x + 16}{2}$$

A) $x \geq -32$ B) $x \leq 5$

C) $x \leq -63$ D) $x \geq 5$

85) For which of the following values of x is this inequality is true ?

$$-1 > \frac{-15 + x}{24}$$

A) $x < -9$ B) $x > -9$

C) $x > -57$ D) $x < -57$

86) For which of the following values of n is this inequality is true ?

$$8(n + 13) < -112$$

A) $n < -2$ B) $n < -27$

C) $n > -94$ D) $n < -94$

87) For which of the following values of b is this inequality is true ?

$$-187 < 13 + 20b$$

A) $b > -34$ B) $b < -10$

C) $b > -90$ D) $b > -10$

88) For which of the following values of p is this inequality is true ?

$$-3(18 + p) \geq -123$$

A) $p \leq -11$ B) $p \leq -27$

C) $p \leq 23$ D) $p \leq -51$

89) For which of the following values
of b is this inequality is true ?

$$\frac{6+b}{3} \geq -2$$

A) b ≥ −12   B) b ≥ −58

C) b ≥ −35   D) b ≥ −80

90) For which of the following values
of x is this inequality is true ?

$$4x - 11 \leq 113$$

A) x ≤ 31   B) x ≤ −3

C) x ≥ −71   D) x ≤ −71

91) For which of the following values
of x is this inequality is true ?

$$-25 < -14 + \frac{x}{3}$$

A) x > 15   B) x > −35

C) x < 15   D) x > −33

92) For which of the following values
of r is this inequality is true ?

$$602 > 14 (9 + r)$$

A) r > − 6   B) r > 34

C) r < 34   D) r < − 6

1) Solve the compound inequality and graph its solution. Represent the solution in :
   Set builder notation and interval notation .

   $-4x < -40$ or $\dfrac{x}{4} \leq 2$

   A) $x \geq 7$ or $x < 6$:

   B) $x \leq 8$:

   C) $x > -7$ or $x \leq -14$:

   D) $x > 10$ or $x \leq 8$:

2) Solve the compound inequality and graph its solution. Represent the solution in :
   Set builder notation and interval notation .

   $4 \leq a + 6 < 16$

   A) $-5 < a \leq 12$:

   B) $a \geq 2$:

   C) $-8 < a \leq 13$:

   D) $-2 \leq a < 10$:

3) Solve the compound inequality and graph its solution. Represent the solution in :
   Set builder notation and interval notation .

   $-11 + m < -4$ and $\dfrac{m}{5} > -1$

   A) $-13 \leq m < 10$: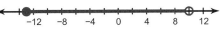

   B) { All real numbers. } :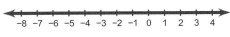

   C) $m < 7$:

   D) $-5 < m < 7$:

4) Solve the compound inequality and graph its solution. Represent the solution in :
   Set builder notation and interval notation .

   $6x \leq -42$ or $x + 9 > 10$

   A) $x \leq -7$ or $x > 1$: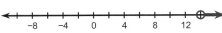

   B) $x > 14$: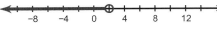

   C) $x < 2$:

   D) $x < -11$:

www.math-knots.com | www.a4ace.com

5) Solve the compound inequality and graph its solution. Represent the solution in :
Set builder notation and interval notation .

$5n > -45$ or $n + 6 \leq -4$

A) $n > -9$ or $n \leq -10$:

B) $n \leq -6$: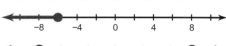

C) $n \geq 14$ or $n \leq -14$:

D) $n \leq -10$:

6) Solve the compound inequality and graph its solution. Represent the solution in :
Set builder notation and interval notation .

$\dfrac{x}{3} \leq 1$ and $-12 + x \geq -11$

A) $x \leq 3$: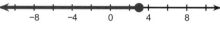

B) $-14 < x \leq -4$:

C) $1 \leq x \leq 3$: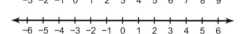

D) No solution.:

7) Solve the compound inequality and graph its solution. Represent the solution in :
Set builder notation and interval notation .

$-10b > -30$ and $b + 12 \geq 12$

A) No solution.: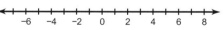

B) $b < 3$:

C) $10 \leq b < 12$:

D) $0 \leq b < 3$:

8) Solve the compound inequality and graph its solution. Represent the solution in :
Set builder notation and interval notation .

$13 + r > 27$ or $-2r > -6$

A) $r \geq 11$: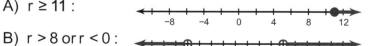

B) $r > 8$ or $r < 0$:

C) $r \leq -14$: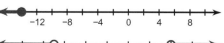

D) $r > 14$ or $r < 3$:

www.math-knots.com | www.a4ace.com

9) Solve the compound inequality and graph its solution. Represent the solution in :
    Set builder notation and interval notation .

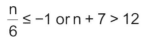 $\dfrac{n}{6} \le -1$ or $n + 7 > 12$

A) $n \le -11$:

B) $n \ge 13$: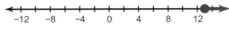

C) $n \le -6$ or $n > 5$: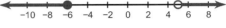

D) $n \le -10$: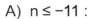

10) Solve the compound inequality and graph its solution. Represent the solution in :
    Set builder notation and interval notation .

$v - 8 \ge -8$ or $\dfrac{v}{7} < -1$

A) $v \le -5$ or $v \ge 8$: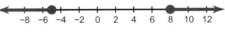

B) { All real numbers. }:

C) $v \ge 0$ or $v < -7$:

D) $v < -10$: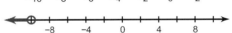

11) Solve the compound inequality and graph its solution. Represent the solution in :
    Set builder notation and interval notation .

$7 \ge k + 12 > -2$

A) $-1 < k \le 10$: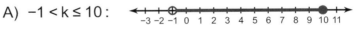

B) $-5 \le k < 8$:

C) $-14 < k \le -5$:

D) $-10 \le k < -9$:

12) Solve the compound inequality and graph its solution. Represent the solution in :
    Set builder notation and interval notation .

$x - 4 \le -10$ or $\dfrac{x}{8} > 1$

A) $x < 2$:

B) $x \le -6$ or $x > 8$: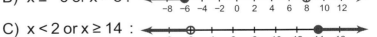

C) $x < 2$ or $x \ge 14$:

D) $x \ge 7$: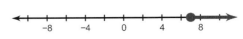

               www.math-knots.com | www.a4ace.com

13) Solve the compound inequality and graph its solution. Represent the solution in :
Set builder notation and interval notation .

$-6 + x < 7$ and $x - 3 > -6$

A) $-14 < x \le 5$:

B) $-3 < x < 13$:

C) $-10 < x \le 6$:

D) $x < 13$:

14) Solve the compound inequality and graph its solution. Represent the solution in :
Set builder notation and interval notation .

$36 \le 12k \le 72$

A) No solution.:

B) $-1 < k < 11$:

C) $k \le 6$:

D) $3 \le k \le 6$:

15) Solve the compound inequality and graph its solution. Represent the solution in :
Set builder notation and interval notation .

$-6m > -12$ or $-11m \le -77$

A) $m < 2$ or $m \ge 7$:

B) $m < 0$:

C) $m < 0$ or $m \ge 12$:

D) $m \ge 7$:

16) Solve the compound inequality and graph its solution. Represent the solution in :
Set builder notation and interval notation .

$x - 1 \ge -3$ or $10 + x \le 5$

A) $x < -7$:
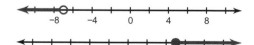

B) $x \ge 5$:

C) $x \ge -2$ or $x \le -5$:

D) $x \ge -2$:

www.math-knots.com | www.a4ace.com

Algebra 1

17) Solve the compound inequality and graph its solution. Represent the solution in :
Set builder notation and interval notation .

$k - 7 \geq -8$ or $\dfrac{k}{14} \leq -1$

A) No solution.: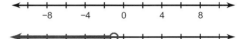

B) $k < -1$:

C) $k \geq -1$:

D) $k \geq -1$ or $k \leq -14$:

18) Solve the compound inequality and graph its solution. Represent the solution in :
Set builder notation and interval notation .

$n + 9 \geq 17$ or $n + 14 < 3$

A) $n \geq -10$: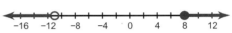

B) $n > 1$: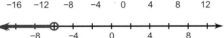

C) $n \geq 8$ or $n < -11$:

D) $n < -6$:

19) Solve the compound inequality and graph its solution. Represent the solution in :
Set builder notation and interval notation .

$n + 10 < 14$ and $n + 7 > 7$

A) $0 < n < 4$: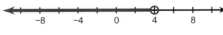

B) $-2 \leq n < 4$:

C) $n < 4$:

D) $n \geq -2$:

20) Solve the compound inequality and graph its solution. Represent the solution in :
Set builder notation and interval notation .

$-10 \leq 2 + a < 8$

A) $-2 < a < -1$:

B) $a < 6$:

C) $-12 \leq a < 6$:

D) $-8 \leq a \leq -6$: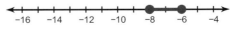

www.math-knots.com | www.a4ace.com

21) Solve the compound inequality and graph its solution. Represent the solution in :
   Set builder notation and interval notation .

$\dfrac{x}{10} < 1$ and $-11x \le 154$

A) $x < 10$:

B) $-14 \le x < 10$:

C) $x \ge -11$:

D) $x > -5$:

22) Solve the compound inequality and graph its solution. Represent the solution in :
   Set builder notation and interval notation .

$x + 5 < 8$ and $x - 10 > -11$

A) $-10 \le x \le 0$:

B) $-11 < x < 10$:

C) $-1 < x < 3$:

D) $x < 3$:

23) Solve the compound inequality and graph its solution. Represent the solution in :
   Set builder notation and interval notation .

$\dfrac{r}{6} \ge -2$ and $\dfrac{r}{7} \le 1$

A) $-12 \le r \le 7$:

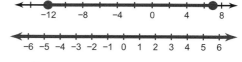

B) $-5 < r < 8$:

C) $r \ge -12$:

D) { All real numbers. }:

24) Solve the compound inequality and graph its solution. Represent the solution in :
   Set builder notation and interval notation .

$5 + x < 4$ or $\dfrac{x}{9} \ge 0$

A) $x \le -14$ or $x \ge -12$:

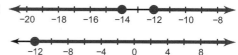

B) $x \ge -12$:

C) No solution.:

D) $x < -1$ or $x \ge 0$:

   www.math-knots.com | www.a4ace.com

25) Solve the compound inequality and graph its solution. Represent the solution in :
Set builder notation and interval notation .

$\dfrac{k}{14} \geq 0$ or $12 + k < 4$

A) $k \geq 0$ or $k < -8$:

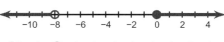

B) $k < -6$:

C) { All real numbers. }:

D) $k \leq -6$:

26) Solve the compound inequality and graph its solution. Represent the solution in :
Set builder notation and interval notation .

$\dfrac{n}{14} > -1$ and $n - 4 < 1$

A) $n \geq 8$:

B) $n > -14$:

C) $n < 5$:

D) $-14 < n < 5$:

27) Solve the compound inequality and graph its solution. Represent the solution in :
Set builder notation and interval notation .

$n - 4 \geq -5$ or $\dfrac{n}{4} \leq -1$

A) No solution.:

B) $n > 8$:

C) $n \geq -1$ or $n \leq -4$:

D) $n > -4$:

28) Solve the compound inequality and graph its solution. Represent the solution in :
Set builder notation and interval notation .

$x - 8 \leq 4$ and $\dfrac{x}{5} > 0$

A) $x < -1$:

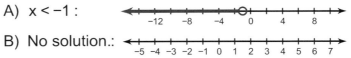

B) No solution.:

C) $x \geq -13$:

D) $0 < x \leq 12$:

www.math-knots.com | www.a4ace.com

29) Solve the compound inequality and graph its solution. Represent the solution in :
Set builder notation and interval notation .

$-169 < 13m \leq -143$

A) $-5 \leq m < 10$: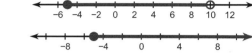

B) $m \geq -5$:

C) $-13 < m \leq -11$: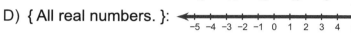

D) { All real numbers. }: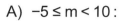

30) Solve the compound inequality and graph its solution. Represent the solution in :
Set builder notation and interval notation .

$-10 + n > -24$ and $13n < -130$

A) $-9 \leq n \leq -5$: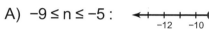

B) $-14 < n < -10$:

C) $2 < n \leq 6$:

D) No solution.:

31) Solve the compound inequality.

$n - 11 \geq -15$ or $13n < -91$

A) No solution.

B) $n \leq -9$ or $n \geq -3$

C) $n < 2$

D) $n \geq -4$ or $n < -7$

32) Solve the compound inequality.

$\frac{x}{2} > 1$ or $x + 6 \leq 7$

A) $x > -4$ or $x < -6$

B) $x > 2$

C) $x > 2$ or $x \leq 1$

D) $x \geq 5$

www.math-knots.com | www.a4ace.com

33) Solve the compound inequality.

$r + 5 < 18$ and $3r > -36$

A) $r > -9$

B) $-12 < r < 13$

C) $r < 14$

D) $-9 < r \leq 4$

34) Solve the compound inequality.

$v + 12 \geq 21$ or $\dfrac{v}{7} \leq -1$

A) $v \leq -7$

B) $v < 1$ or $v \geq 6$

C) $v \leq 5$ or $v > 6$

D) $v \geq 9$ or $v \leq -7$

35) Solve the compound inequality.

$-9v \leq 99$ and $13v \leq -52$

A) $-6 < v < 0$

B) $-11 \leq v \leq -4$

C) $v \geq -11$

D) $v \leq -4$

36) Solve the compound inequality.

$11x - 10 \leq 100$ and $7x - 5 > -68$

A) $x \leq 13$

B) $10 < x \leq 13$

C) $x > 10$

D) $-9 < x \leq 10$

37) Solve the compound inequality.

$11n - 4 > -114$ or $14 + 5n < -56$

A) $n > -5$ or $n < -8$

B) $n \leq -4$

C) $n > -10$ or $n < -14$

D) $n \leq -9$ or $n > 10$

38) Solve the compound inequality.

$5 + 13k \geq -8$ or $-2 + 10k < -142$

A) $k \geq -1$

B) { All real numbers. }

C) $k \geq -1$ or $k < -14$

D) $k \leq -9$

39) Solve the compound inequality.

$-21 \leq 7r + 7 \leq 42$

A) $-10 \leq r \leq 3$

B) $r < 3$

C) $-4 \leq r \leq 5$

D) $r \leq 3$

40) Solve the compound inequality.

$14m + 13 \leq -183$ or $12m + 2 \geq 14$

A) $m \leq -8$ or $m > -1$

B) No solution.

C) $m \leq -14$ or $m \geq 1$

D) $m \geq 13$

 **Algebra 1**

41) Solve the compound inequality.

$2b - 8 \geq -34$ and $-b - 2 > -4$

A) $0 \leq b \leq 9$

B) $0 < b \leq 4$

C) $b > -11$

D) $-13 \leq b < 2$

42) Solve the compound inequality.

$6 \leq -5 + 11x < 50$

A) $1 \leq x < 5$

B) $x \leq 10$

C) $x < 6$

D) $9 \leq x \leq 10$

43) Solve the compound inequality.

$7a - 7 > 35$ or $-4a + 4 \geq 40$

A) $a > 6$ or $a \leq -9$

B) $a > -9$

C) $a \leq -2$

D) $a > 11$

44) Solve the compound inequality.

$-148 < 6 - 14x < 76$

A) $-5 < x < 11$

B) $-11 \leq x \leq -3$

C) $x < 1$

D) $1 \leq x \leq 9$

45) Solve the compound inequality.

$-3n - 5 > 19$ or $-13 + 8n \geq -21$

A) $n < -8$ or $n \geq -1$

B) $n < -8$

C) $n < -12$

D) $n \leq -8$

46) Solve the compound inequality.

$5r + 8 < -12$ or $10r - 11 > 109$

A) $r < -4$ or $r > 12$

B) $r \geq -8$

C) $r > 2$

D) $r < -7$

47) Solve the compound inequality.

$-8x - 1 < -41$ and $8x + 11 < 115$

A) $x < 13$

B) $x > -4$

C) $-4 < x < 13$

D) $5 < x < 13$

48) Solve the compound inequality.

$13x - 11 \geq 15$ or $11x - 11 < -143$

A) $x > -4$ or $x \leq -6$

B) $x > 12$ or $x < 2$

C) $x \geq 9$

D) $x \geq 2$ or $x < -12$

                 www.math-knots.com | www.a4ace.com

49) Solve the compound inequality.

$7 + 13b \leq -110$ and $12b - 7 \geq -175$

A) $b \leq -9$

B) $b < 11$

C) $-14 \leq b \leq -9$

D) No solution.

50) Solve the compound inequality.

$-8 \leq -13 + x \leq -7$

A) No solution.

B) $5 \leq x \leq 6$

C) $-9 \leq x \leq 6$

D) $8 < x < 12$

51) Solve the compound inequality.

$x - 6 \geq -15$ and $12x + 10 \leq 178$

A) $-9 \leq x \leq 14$

B) $1 \leq x < 14$

C) $-12 < x < 0$

D) $x \leq 14$

52) Solve the compound inequality.

$-14x - 2 \geq -142$ and $6x - 13 \geq -85$

A) $-5 < x < 3$

B) $x < 3$

C) $-12 \leq x \leq 10$

D) $-14 \leq x \leq -7$

53) Solve the compound inequality.

$-135 < 11v + 8 < -47$

A) $v < 8$

B) $0 < v \leq 13$

C) $-13 < v < -5$

D) $v \leq 13$

54) Solve the compound inequality.

$1 - 12x < -59$ or $5x + 2 < -18$

A) $x \geq 12$

B) $x > 5$ or $x < -4$

C) $x < 0$ or $x \geq 12$

D) $x < -7$ or $x > 7$

55) Solve the compound inequality.

$-9 + 6n > -69$ or $7 + 10n < -133$

A) $n > -10$ or $n < -14$

B) $n < -14$

C) No solution.

D) $n > -10$

56) Solve the compound inequality.

$-2 \leq -14 - 12p \leq 142$

A) $-13 \leq p \leq -1$

B) $-7 \leq p < 7$

C) $p < 7$

D) $p \leq -1$

57) Solve the compound inequality.

$9 - 6n \geq -45$ and $-13n + 2 < 2$

A) { All real numbers. }

B) $n \geq -4$

C) $0 < n \leq 9$

D) $-14 < n \leq -11$

58) Solve the compound inequality.

$6n + 8 > 50$ and $12n + 10 \leq 154$

A) $n \leq 13$

B) $7 < n \leq 12$

C) $n > 7$

D) No solution.

59) Solve the compound inequality.

$55 > 5 + 10n > -125$

A) $-14 \leq n < 6$

B) $-13 < n < 5$

C) $n < 6$

D) $n \leq -11$

60) Solve the compound inequality.

$-10k + 10 \leq -40$ or $6 + 3k \leq 0$

A) $k \geq -1$

B) $k \geq 5$ or $k \leq -2$

C) $k \geq -1$ or $k < -5$

D) $k \geq 2$ or $k < -9$

61) Solve the compound inequality.

$105 \leq 10b - 5 < 125$

A) $b \leq -3$

B) $b \geq 11$

C) $-10 < b < 10$

D) $11 \leq b < 13$

62) Solve the compound inequality.

$7 - 11m \geq 29$ and $9m + 1 > -89$

A) $m < 5$

B) $m > -2$

C) $-2 < m \leq 6$

D) $-10 < m \leq -2$

63) Solve the compound inequality.

$-134 < -14n - 8 \leq 146$

A) $n < 9$

B) $-11 \leq n < 9$

C) $-13 \leq n < 0$

D) No solution.

64) Solve the compound inequality.

$-35 \leq 3n - 11 < -17$

A) $-8 \leq n < -2$

B) $10 < n \leq 11$

C) $n > -1$

D) $6 < n \leq 9$

65) Solve the compound inequality.

$$28 < -3n + 13 \le 31$$

A) $n < 6$

B) $-6 \le n < -5$

C) $n > -1$

D) $-1 < n < 6$

66) Solve the compound inequality.

$$n + 0.7 > -3.9 \text{ and } -4.3n > 15.91$$

A) $-4.6 < n < -3.7$

B) $n < 2.2$

C) $-7.1 \le n < 2.2$

D) $n < -3.8$

67) Solve the compound inequality.

$$-15.6 < 7.8x \le 14.04$$

A) $1.92 \le x < 6.5$

B) $x > 0.13$

C) $x \le 8$

D) $-2 < x \le 1.8$

68) Solve the compound inequality.

$$-3.1m > -24.8 \text{ and } -1.6m \le -4.64$$

A) $2.9 \le m < 8$

B) $m \ge -8.6$

C) $m \ge 4.2$

D) $-8.6 \le m \le 3.7$

69) Solve the compound inequality.

$$\frac{n}{10} < -\frac{9}{10} \text{ or } n + 8.8 \ge 10.6$$

A) $n < -9 \text{ or } n \ge 1.8$

B) $n \ge 1.8$

C) $n > -2.3 \text{ or } n < -5.1$

D) $n \le -2.6$

70) Solve the compound inequality.

$$-3.8 \le n - 6.5 \le 1.8$$

A) $n > -0.2$

B) $n < 9.5$

C) $n \le 6.3$

D) $2.7 \le n \le 8.3$

71) Solve the compound inequality.

$$-17.4 < 3n < -1.5$$

A) $-0.7 \le n < 7.6$

B) $-5.8 < n < -0.5$

C) $n < -1.4$

D) $n \ge -0.7$

72) Solve the compound inequality.

$$-0.2 \le -3 + k < 2$$

A) $-2.2 \le k < 1.3$

B) $k \ge 2.8$

C) No solution.

D) $2.8 \le k < 5$

73) Solve the compound inequality.

$$2.91 \geq -0.3a > 0.03$$

A) No solution.

B) $-6.9 \leq a \leq -2.7$

C) $-9.7 \leq a < -0.1$

D) $2.4 \leq a \leq 3.4$

74) Solve the compound inequality.

$$a - 4.9 < -9 \quad \text{or} \quad a - 6.8 > -4.1$$

A) $a \leq 3.1 \text{ or } a \geq 7.75$

B) $a > 8.5 \text{ or } a < -1.7$

C) $a > 8.5$

D) $a < -4.1 \text{ or } a > 2.7$

75) Solve the compound inequality.

$$1.2 \leq r + 8.5 < 18.2$$

A) $-3 \leq r < 3.8$

B) $r > 6.78$

C) $6.78 < r \leq 9.6$

D) $-7.3 \leq r < 9.7$

76) Solve the compound inequality.

$$x + 0.4 \geq -7.1 \quad \text{and} \quad 9.7 + x \leq 16.5$$

A) $x \leq 6.8$

B) $2.3 < x < 5.9$

C) $x \geq 5.965$

D) $-7.5 \leq x \leq 6.8$

77) Solve the compound inequality.

$$v + 6.91 > 16.31 \quad \text{or} \quad 9.7v < 49.47$$

A) $v > 9.4$

B) $v < 5.1$

C) $v < 1.3 \text{ or } v \geq 3.7$

D) $v > 9.4 \text{ or } v < 5.1$

78) Solve the compound inequality.

$$-6.8x \leq -30.6 \quad \text{or} \quad 5.1x \leq -8.67$$

A) $x \geq 4.5$

B) $x \leq -1.7$

C) $x < -5.9 \text{ or } x > 3.7$

D) $x \geq 4.5 \text{ or } x \leq -1.7$

79) Solve the compound inequality.

$$r - 8.5 > -2.8 \text{ or } -2.7 + r < -12.6$$

A) $r \leq -4.3$

B) $r > 5.7 \text{ or } r < -9.9$

C) $r \geq 5.4 \text{ or } r < 5.2$

D) $r \geq 6.234$

80) Solve the compound inequality.

$$b + 8.7 > 9.8 \quad \text{and} \quad b - 6.1 \leq -4.2$$

A) $1.1 < b \leq 1.9$

B) $3.4 \leq b < 9.24$

C) $-1.67 < b < 4.9$

D) $b \geq -3.8$

81) Solve the compound inequality.

$$13 > r + 7.2 > 10.4$$

A) $3.2 < r < 5.8$

B) $-8.12 < r < 2.7$

C) $r > 3.2$

D) $-4 \le r \le 4$

82) Solve the compound inequality.

$$-2.1 \le r + 1.3 < 1$$

A) $r \ge -3.4$

B) $-3.4 \le r < -0.3$

C) $0.4 \le r \le 7.2$

D) $-9.3 \le r \le 7.4$

83) Solve the compound inequality.

$$x - 7.6 \ge -14.2 \ \text{and} \ x + 4.5 < 9.8$$

A) $x \ge -6.6$

B) $-6.6 \le x < 5.3$

C) $x < 5.3$

D) $x \le 6.6$

84) Solve the compound inequality.

$$-0.3 \le 6 + v \le 14.9$$

A) $-4.8 \le v \le 6.8$

B) $v \le 6.8$

C) $v \le -5.3$

D) $-6.3 \le v \le 8.9$

85) Solve the compound inequality.

$$5.8 < p + 8.5 < 16.9$$

A) $-4.71 \le p < -1$

B) { All real numbers. }

C) $p < 8.4$

D) $-2.7 < p < 8.4$

86) Solve the compound inequality.

$$3 < x + 8.2 < 15.2$$

A) $x < 7$

B) $x \ge -4.7$

C) $-5.2 < x < 7$

D) $x > -5.2$

87) Solve the compound inequality.

$$-3.7 + n \ge -8.51 \ \text{or} \ n + 5.19 \le -4.31$$

A) $n < -4.9$

B) $n > 6.8 \ \text{or} \ n < -5.6$

C) $n \ge -4.81 \ \text{or} \ n \le -9.5$

D) { All real numbers. }

88) Solve the compound inequality.

$$-3.1 < -0.1 + v \le 5.8$$

A) { All real numbers. }

B) $v \le 7.5$

C) $-2.9 < v < 1.1$

D) $-3 < v \le 5.9$

89) Solve the compound inequality.

$$-23.49 < 2.9r \leq -17.11$$

A) $r \leq -5.9$

B) $r \geq -7.87$

C) $-8.1 < r \leq -5.9$

D) $-5.2 \leq r < -5$

90) Solve the compound inequality.

$$x - 2.7 \leq -12.1 \text{ or } 7.09 + x \geq 13.59$$

A) $x \leq -9.4 \text{ or } x \geq 6.5$

B) $x \leq -9.4$

C) $x < 4$

D) $x > 4.4 \text{ or } x < 4$

91) Solve the compound inequality.

$$1.2k \leq -7.92 \text{ or } k + 2.8 \geq -3.2$$

A) $k \leq -1 \text{ or } k > 2$

B) No solution.

C) $k \leq -6.6 \text{ or } k \geq -6$

D) $k < -3.2 \text{ or } k > 0.8$

92) Solve the compound inequality.

$$x + 8.9 < 13.4 \text{ or } 2.8x > 14.56$$

A) $x < 4.5 \text{ or } x > 5.2$

B) $x \geq 5$

C) $x \leq 1.5 \text{ or } x > 9.45$

D) $x \geq 5 \text{ or } x < -7.1$

93) Solve the compound inequality.

$$7.2a > 65.52 \text{ or } a - 8.5 \leq -16$$

A) $a > 9.1 \text{ or } a \leq -7.5$

B) $a \geq 6.2$

C) $a \leq -8.9$

D) $a < -8 \text{ or } a > -3.1$

94) Solve the compound inequality.

$$9.5 \leq -9.5n \leq 78.85$$

A) $-8.3 \leq n \leq -1$

B) $n \geq -8.3$

C) $n \geq 3.9$

D) $3.9 \leq n \leq 6.8$

95) Solve the compound inequality.

$$\frac{k}{0.6} \geq 5.5 \text{ or } -2.7 + k \leq -10.7$$

A) $k \geq 8.5 \text{ or } k < -7.1$

B) $k \leq -8$

C) $k \leq 5.25 \text{ or } k > 7.5$

D) $k \geq 3.3 \text{ or } k \leq -8$

96) Solve the compound inequality.

$$x + 0.1 \geq 5.9 \text{ or } x - 0.7 \leq -2.9$$

A) $x \leq 2.8$

B) $x < -2.4$

C) $x \geq 5.8 \text{ or } x \leq -2.2$

D) $x > 3.38$

97)   Solve the compound inequality.

$-19.5 < -5a \le 12.5$

A) $-8.5 \le a < 5.1$

B) $a < 3.9$

C) $a \ge -8.5$

D) $-2.5 \le a < 3.9$

98)   Solve the compound inequality.

$x + 5.6 > 11.32 \ \text{ or } \ -4x > -11.8$

A) $x \le -6.6$

B) $x > -4.5$

C) $x > 5.72$

D) $x > 5.72 \ \text{or} \ x < 2.95$

99)   Solve the compound inequality.

$-8.5x + 1.4 \ge 52.4 \ \text{ or } \ 4.8x - 1.9 > -23.5$

A) $x \le -6$

B) $x < -3 \ \text{or} \ x > 1.4$

C) $x \le -6 \ \text{or} \ x > -4.5$

D) $x > -4.5$

100)   Solve the compound inequality.

$0.9n + 4.4 \ge 3.23 \ \text{ and } \ -5 + 0.4n \le -2.04$

A) No solution.

B) $n \le 8.7$

C) $n > -6.1$

D) $-1.3 \le n \le 7.4$

101)   Solve the compound inequality.

$-15.2 < -7.3 - v \le -11.86$

A) $v \le 6.4$

B) $v \ge 4.9$

C) $4.56 \le v < 7.9$

D) $4.9 \le v \le 6.4$

102)   Solve the compound inequality.

$-5.5 < 2.5r + 6 < 8$

A) $-4.6 < r < 0.8$

B) $-5.8 < r \le 0.4$

C) $r < 0.8$

D) $-4.1 \le r \le -2.6$

103)   Solve the compound inequality.

$0.6r - 2.7 < -8.34 \ \text{ or } \ 2.9r - 0.5 \ge -23.99$

A) $r \ge -0.1$

B) $r \ge -3.1$

C) $r \ge -8.1$

D) $r < -9.4 \ \text{or} \ r \ge -8.1$

104)   Solve the compound inequality.

$-15.6 < -3.6n + 6 < 17.52$

A) $n \le 1.6$

B) $6.39 < n < 9.02$

C) No solution.

D) $-3.2 < n < 6$

105) Solve the compound inequality.

$0.3 - 1.6v < -8.66$ or $-9.6 - 5.4v > -19.86$

A) $v < 1.9$

B) $v > 5.6$ or $v < 1.9$

C) $v > -0.6$

D) $v \leq -4.2$

106) Solve the compound inequality.

$-7.7n + 7.1 > 81.02$ or $0.48 - 0.2n < 2.126$

A) $n \leq -8.01$

B) $n \geq 7.4$ or $n \leq -8.01$

C) $n < -9.6$ or $n > -8.23$

D) { All real numbers. }

107) Solve the compound inequality.

$49.62 < -7.7x + 6.5 < 81.19$

A) $-9.7 < x < -5.6$

B) $x > -5.8$

C) $-9.4 \leq x \leq 4.1$

D) $-4 \leq x \leq 6.9$

108) Solve the compound inequality.

$27.57 < 6.1x - 7.2 \leq 28.79$

A) $4.4 \leq x \leq 10$

B) $x \leq 5.9$

C) $5.7 < x \leq 5.9$

D) $x > 5.7$

109) Solve the compound inequality.

$9.3 + 4.9r \leq -11.28$ or $8.3r + 6.3 > -5.32$

A) $r \leq -4.2$ or $r > -1.4$

B) $r > -1.4$

C) $r \leq -4.2$

D) $r < -6.42$

110 Solve the compound inequality.

$7.6 + 7.3b \leq -40.58$ or $-6.5b - 7.4 \leq -33.4$

A) $b < -7.5$ or $b > 2.5$

B) $b > 9.8$ or $b \leq 4.6$

C) $b \leq -6.6$ or $b \geq 4$

D) No solution.:

111 Solve the compound inequality.

$6.7b + 4.6 \geq -49$ and $6.2 + 4.7b \leq 48.03$

A) $-8 \leq b \leq 8.9$

B) $-4.6 \leq b \leq -3.69$

C) $-8.5 \leq b \leq -5.4$

D) $b \leq -3.69$

112 Solve the compound inequality.

$1.5x - 9.3 < -22.95$ or $-4.4x + 2.2 < 33.88$

A) $x \leq -7.7$ or $x \geq -4.6$

B) { All real numbers. }

C) $x \geq -4.6$

D) $x < -9.1$ or $x > -7.2$

113) Solve the compound inequality

$-19.86 < -5.8 + 3.8n \le -1.62$

A) $n < 7$

B) $-6.2 \le n < 7$

C) $-3.7 < n \le 1.1$

D) $n \le \dfrac{13}{2}$

114) Solve the compound inequality.

$7.2x - 3.9 > 6.18$ or $4.9x - 3.9 < -36.24$

A) $x > 1.4$ or $x < -6.6$

B) $x \ge 8.8$

C) $x > 9.7$ or $x \le 4.8$

D) $x \le 6.7$

115) Solve the compound inequality.

$-0.2 - 8.2p \ge -29.72$ or $6.5p - 5.9 \ge 35.7$

A) $p \ge 8$

B) $p \le 3.6$ or $p \ge 6.4$

C) $p \le 6$ or $p \ge 8$

D) $p \ge 6.4$

116) Solve the compound inequality.

$6.6n - 7.1 > 52.3$ or $1.7n - 0.3 < -14.07$

A) $n < -6.6$

B) $n > 9$ or $n < -8.1$

C) $n > 8.3$ or $n < -5.8$

D) $n \ge 9.8$ or $n \le 1$

117) Solve the compound inequality.

$4.9a - 3.3 \ge 13.36$ or $-4.7a - 5.2 \ge 30.05$

A) $a \ge 8.6$

B) $a \ge 3.4$ or $a \le -7.5$

C) $a < -7.2$

D) $a \ge 3.4$

118) Solve the compound inequality.

$7.6 - 8.3m \le 11.75$ or $1.4m - 8 \le -12.62$

A) $m < -6.3$

B) $m \ge -0.5$ or $m \le -3.3$

C) $m \ge -\dfrac{1}{2}$

D) $m \le -3.3$

119

Solve the compound inequality.

$1.7 - 3.6r > -2.26$ and $5.5r - 7.05 > -34.55$

A) $-5.9 \le r \le -3.4$

B) $-5 < r < 1.1$

C) $-8 < r \le -3$

D) $-6.1 \le r < -2.1$

120) Solve the compound inequality.
$-28.92 < 0.9 + 7.1x \le 4.45$

A) $-4.72 \le x < 0.8$

B) $x \ge -4.72$

C) $-4.2 < x \le 0.5$

D) $-3 < x \le 4.1$

121)   Solve the compound inequality.

1.4x − 1.2 ≤ −13.1 or −7.8x + 7.2 < −29.46

   A) x ≥ 2.4

   B) x ≤ −8.5

   C) x ≤ −8.5 or x > 4.7

   D) x < −3

122)   Solve the compound inequality.

−56.38 ≤ 7.6n − 0.9 ≤ −54.86

   A) −4.83 ≤ n ≤ −4.5

   B) n < 3.08

   C) −7.3 ≤ n ≤ −7.1

   D) { All real numbers. }

123)   Solve the compound inequality.

59.3 ≤ 8.5v + 3.2 < 72.05

   A) v ≤ 3.7

   B) 2.4 < v ≤ 3.7

   C) −6.44 < v < 8.9

   D) 6.6 ≤ v < 8.1

124)   Solve the compound inequality.

−11.44 < −2.2n − 8.8 ≤ 6.82

   A)   n ≥ −9.8
   B)   { All real numbers. }

   C) 1.1 < n < 7

   D) −7.1 ≤ n < 1.2

125) Solve the compound inequality.

7.3x − 0.1 ≤ −18.35 or
0.6 − 4.1x ≤ 6.75

   A) No solution.

   B) { All real numbers. }

   C) x < −2 or x ≥ 9.9

   D) x ≤ −2.5 or x ≥ −1.5

126)   Solve the compound inequality.
2.9x + 7.7 < −7.96 or
6.1x + 4.9 ≥ −24.38

   A) x < −3.2 or x > 9.5

   B) x > 3.3 or x < 2.7

   C) x < −5.4 or x ≥ −4.8

   D) x > 9.5

127)   Solve the compound inequality.
0.9x + 3.02 ≤ −2.47 and
6.4x + 0.2 > −54.2

   A) x > −8.5

   B) 0.9 ≤ x < 6.54

   C) −8.5 < x ≤ −6.1

   D) x < 6.54

128)   Solve the compound inequality.
5.9k + 5.9 > −46.02 and
3.7k − 2.2 < 31.84

   A) { All real numbers. }

   B) k < 9.2

   C) k > −8.8

   D) −8.8 < k < 9.2

129) Solve the compound inequality.
$-57.58 \leq 5.6x - 8.3 < -5.5$

    A) $x < 5.4$

    B) $x \leq 9.88$

    C) $-8.8 \leq x < 0.5$

    D) $0.2 \leq x \leq 9.88$

130) Solve the compound inequality.

$7.8 + 4.4a > 31.56$ or $5.2a - 9.1 \leq -43.42$

    A) $a > 5.4$ or $a \leq -6.6$

    B) $a < 1.1$ or $a > 8.21$

    C) $a \leq -8.61$:

    D) $a > 5.4$:

131) Solve the compound inequality.

$4.2a - 9.3 \geq 10.02$ or $8a - 0.33 < 23.67$

A) $a \geq 4.6$ or $a < 3$

B) $a \leq -3$

C) $a \geq 4.6$

D) $a \leq -9.9$

132) Solve the compound inequality.

$7.54 \leq 1.3 + 3.9n < 14.17$

A) $1.6 \leq n < 3.3$

B) $-3.2 < n \leq -1.7$

C) $n \leq 2.2$

D) $-1.6 \leq n \leq 3.5$

133) Solve the compound inequality.

$-6.68 < -2.2 - 2.8a < 18.52$

    A) $2.8 < a < 7.8$

    B) $-7.4 < a < 1.6$

    C) $a > -7.4$

    D) $a < 7.8$

134) Solve the compound inequality.

$$\frac{29}{6}a \geq \frac{174}{7} \quad \text{or} \quad \frac{14}{9}a < \frac{70}{9}$$

    A) $a \geq \dfrac{36}{7}$ or $a < 5$

    B) $a < 5$

    C) $a \leq \dfrac{2}{7}$

    D) $a \geq \dfrac{3}{8}$

135) Solve the compound inequality.

$$\frac{2}{3}k > \frac{10}{7} \quad \text{or} \quad k - \frac{7}{5} < -\frac{3}{5}$$

    A) $k \geq \dfrac{3}{5}$

    B) $k < -\dfrac{7}{4}$ or $k \geq \dfrac{3}{5}$

    C) $k \geq \dfrac{2}{9}$

    D) $k > \dfrac{15}{7}$ or $k < \dfrac{4}{5}$

136) Solve the compound inequality.

$$-\frac{22}{5}b \ge -\frac{418}{35} \quad \text{or} \quad b - \frac{17}{9} \ge \frac{251}{72}$$

A) $b \le \frac{19}{7}$ or $b \ge \frac{43}{8}$

B) $b < -\frac{5}{4}$

C) { All real numbers. }

D) $b \ge \frac{43}{8}$

137) Solve the compound inequality.

$$a - \frac{4}{3} > 2 \quad \text{or} \quad 1 + a \le -\frac{1}{2}$$

A) $a > \frac{10}{3}$ or $a \le -\frac{3}{2}$

B) $a < -\frac{1}{2}$

C) No solution.

D) $a > \frac{17}{3}$

138) Solve the compound inequality.

$$-\frac{4}{3}x < -\frac{6}{5} \quad \text{and} \quad -\frac{31}{10}x \ge -\frac{589}{50}$$

A) $\frac{9}{10} < x \le \frac{19}{5}$

B) $x \le \frac{19}{5}$

C) $-1 < x < \frac{51}{10}$

D) $x < \frac{51}{10}$

139) Solve the compound inequality

$$-\frac{11}{5} < p + \frac{8}{5} < \frac{259}{40}$$

A) $\frac{26}{7} < p < \frac{31}{6}$

B) $-\frac{19}{5} < p < \frac{39}{8}$

C) $p < \frac{8}{7}$

D) $p > -\frac{19}{5}$

140) Solve the compound inequality.

$$\frac{3}{2}b > -\frac{13}{4} \quad \text{and} \quad 7b \le \frac{147}{8}$$

A) $-\frac{13}{6} < b \le \frac{21}{8}$

B) $b < -\frac{4}{5}$

C) $\frac{2}{3} \le b \le \frac{17}{3}$

D) $-\frac{21}{10} < b \le 2$

1) Which graph represents the linear inequality and its solution.

$$y < \frac{7}{3}x + 4$$

A)
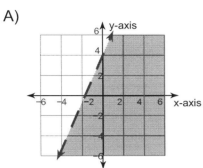

B)
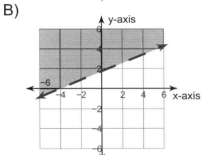

C)
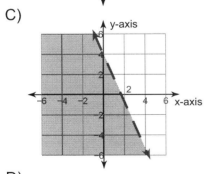

D)
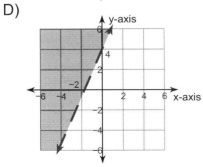

2) Which graph represents the linear inequality and its solution.

$$y < \frac{5}{4}x - 4$$

A)
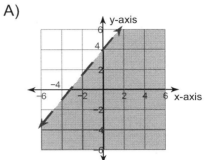

B)
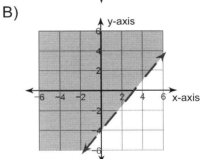

C)
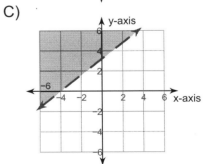

D)
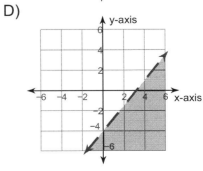

www.math-knots.com | www.a4ace.com

3) Which graph represents the linear inequality and its solution.

$$x < 5$$

A)
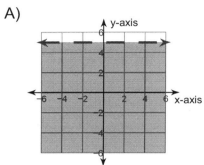

B)
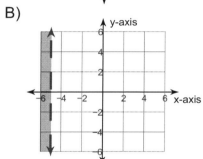

C)
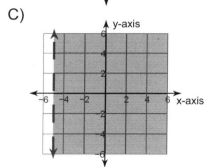

D)
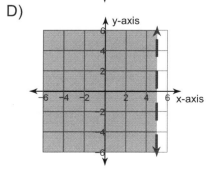

4) Which graph represents the linear inequality and its solution.

$$y \le -\frac{7}{2}x + 2$$

A)
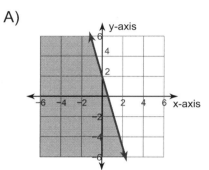

B)
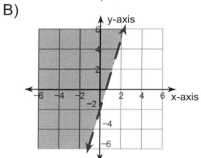

C)
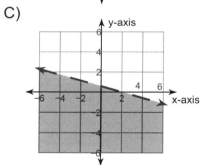

D)
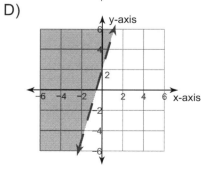

www.math-knots.com | www.a4ace.com

5) Which graph represents the linear inequality and its solution.

$$y > 10x + 5$$

A)
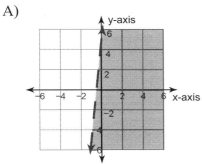

B)
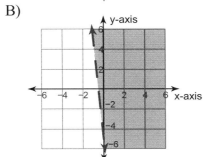

C)
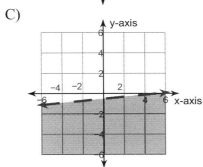

D)
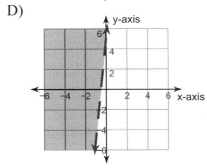

6) Which graph represents the linear inequality and its solution.

$$y > \frac{6}{5}x - 1$$

A)
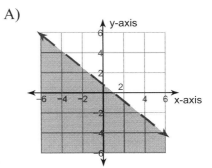

B)
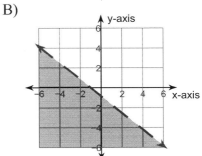

C)
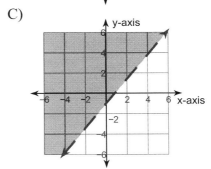

D)
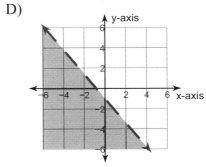

www.math-knots.com | www.a4ace.com

7) Which graph represents the linear inequality and its solution.

$$y \le -8x + 4$$

A)
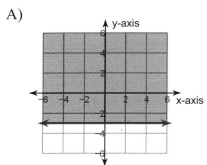

B)
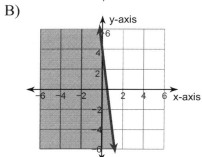

C)
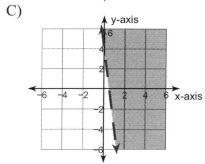

D)
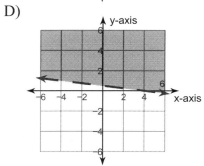

8) Which graph represents the linear inequality and its solution.

$$x > 3$$

A)
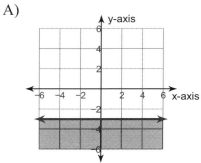

B)
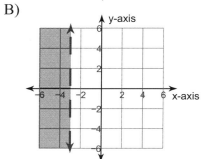

C)
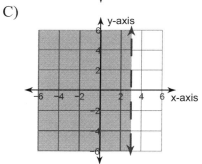

D)
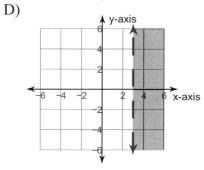

www.math-knots.com | www.a4ace.com

9) Which graph represents the linear inequality and its solution.

$$y \le \frac{1}{3}x + 4$$

A)

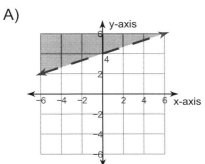

B)

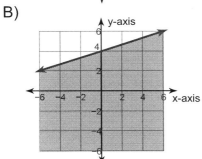

C)

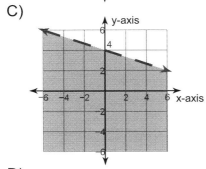

D)

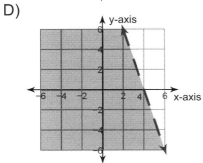

10) Which graph represents the linear inequality and its solution.

$$y > 2x - 4$$

A)

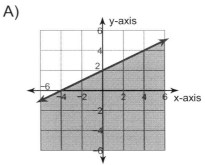

B)

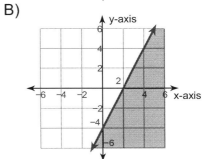

C)

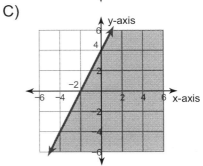

D)

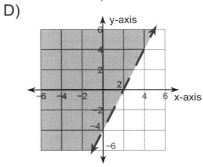

www.math-knots.com | www.a4ace.com

11) Which graph represents the linear inequality and its solution.

$$y \leq -3x - 4$$

A)
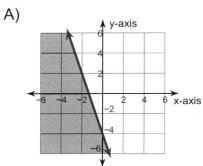

B)
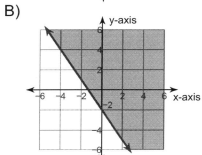

C)
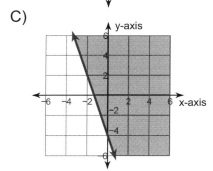

D)
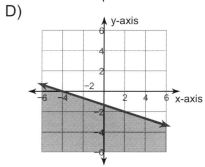

12) Which graph represents the linear inequality and its solution.

$$y < \frac{7}{4}x - 4$$

A)
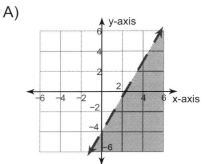

B)
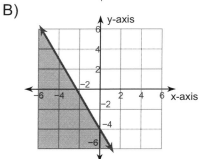

C)
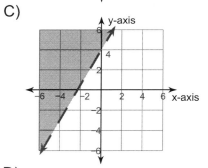

D)
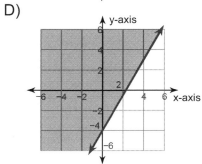

www.math-knots.com | www.a4ace.com

13) Which graph represents the linear inequality and its solution.

$$y < -3x + 2$$

A)

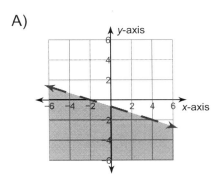

B)

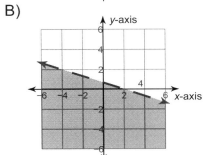

C)

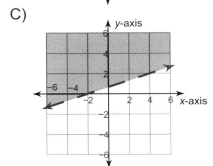

D)

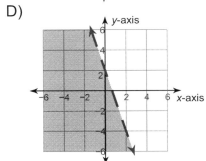

14) Which graph represents the linear inequality and its solution.

$$y < -x - 4$$

A)

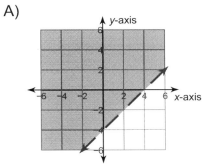

B)

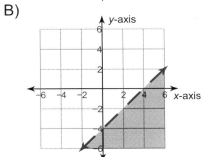

C)

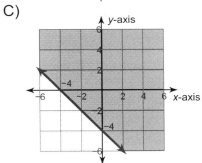

D)

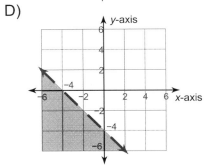

www.math-knots.com | www.a4ace.com

15) Which graph represents the linear inequality and its solution.

$$y \geq \frac{2}{3}x - 5$$

A)

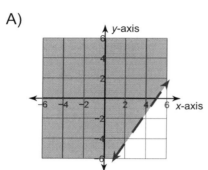

B)

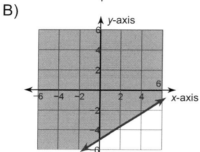

C)

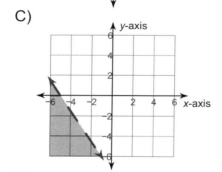

D)

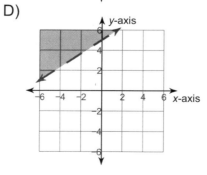

16) Which graph represents the linear inequality and its solution.

$$y \leq -4$$

A)

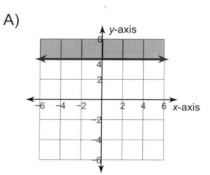

B)

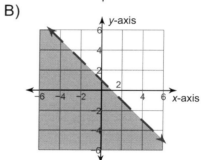

C)

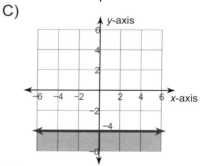

D)

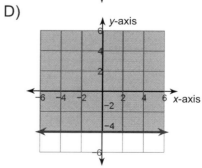

www.math-knots.com | www.a4ace.com

17) Which graph represents the linear inequality and its solution.

$$y < -3x - 4$$

A)
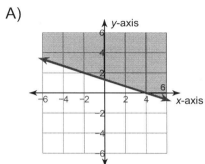

B)
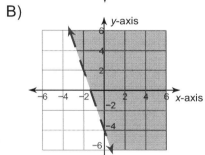

C)
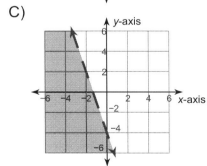

D)
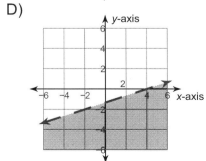

18) Which graph represents the linear inequality and its solution.

$$y < -\frac{3}{2}x - 4$$

A)
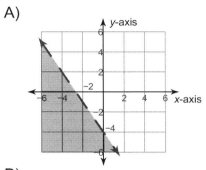

B)
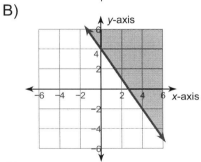

C)
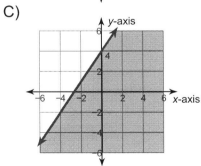

D)
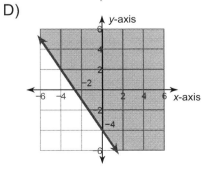

www.math-knots.com | www.a4ace.com

19) Which graph represents the linear inequality and its solution.

$$y < 2x + 2$$

A)
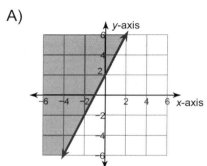

B)
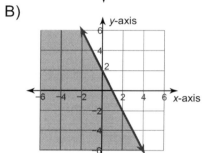

C)
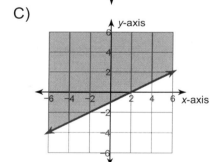

D)
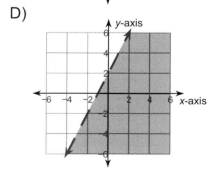

20) Which graph represents the linear inequality and its solution.

$$y \leq \frac{1}{2}x + 2$$

A)
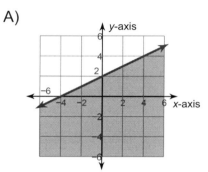

B)
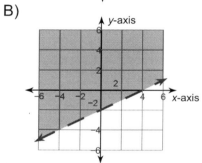

C)
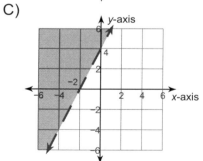

D)
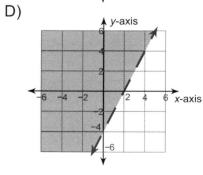

21) Which graph represents the linear inequality and its solution.

$$y > -\frac{1}{2}x - 3$$

A)

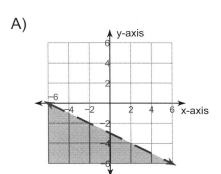

B)

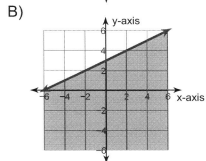

C)

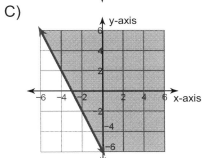

D)

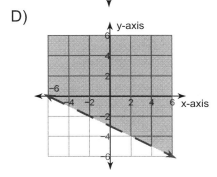

22) Which graph represents the linear inequality and its solution.

$$x \leq -3$$

A)

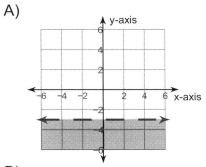

B)

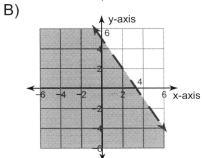

C)

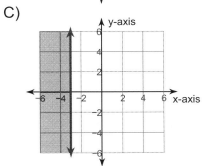

D)

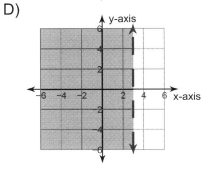

www.math-knots.com | www.a4ace.com

23) Which graph represents the linear inequality and its solution.

$$y \le 3x - 5$$

A)
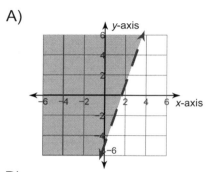

B)
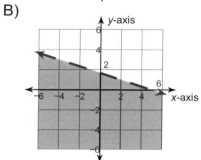

C)
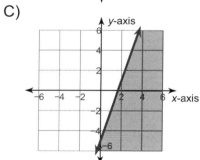

D)
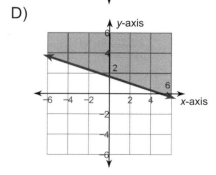

24) Which graph represents the linear inequality and its solution.

$$y > \frac{5}{3}x$$

A)
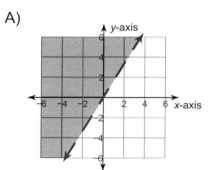

B)
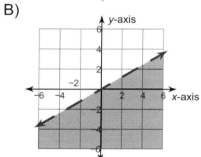

C)
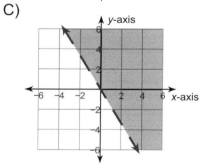

D)
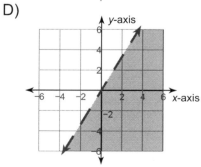

www.math-knots.com | www.a4ace.com

25) Which graph represents the linear inequality and its solution.

$$y \geq -\frac{1}{2}x - 4$$

A)

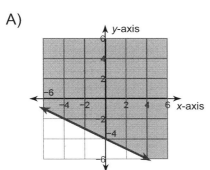

B)

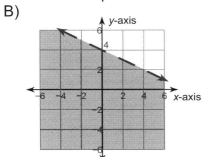

C)

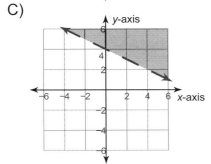

D)

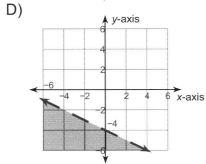

26) Which graph represents the linear inequality and its solution.

$$y > -\frac{1}{2}x - 1$$

A)

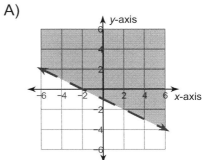

B)

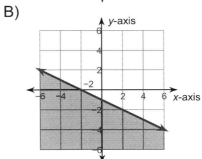

C)

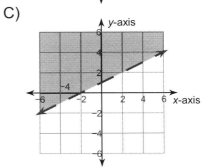

D)

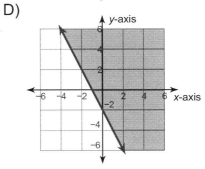

27) Which graph represents the linear inequality and its solution.

$$y \le \frac{4}{5}x + 1$$

A)
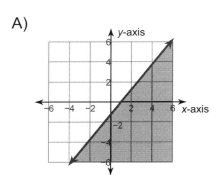

B)
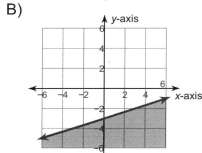

C)
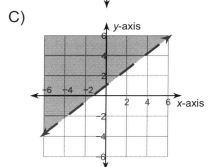

D)
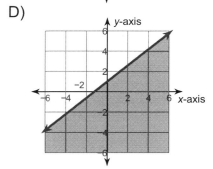

28) Which graph represents the linear inequality and its solution.

$$y \le -\frac{3}{4}x + 3$$

A)
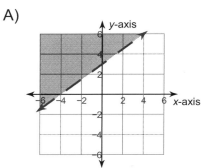

B)
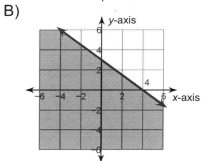

C)
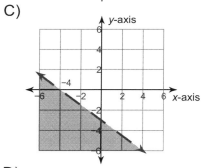

D)
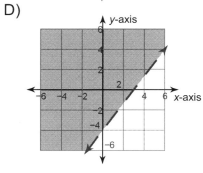

www.math-knots.com | www.a4ace.com

29) Which graph represents the linear inequality and its solution.

$$y > -9x + 4$$

A)

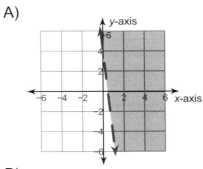

B)

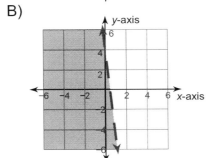

C)

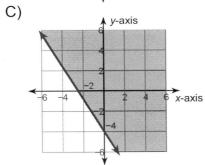

D)

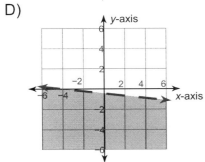

30) Which graph represents the linear inequality and its solution.

$$y \leq \frac{3}{2}x - 3$$

A)

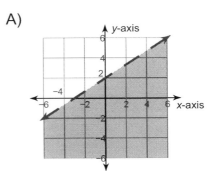

B)

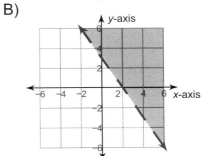

C)

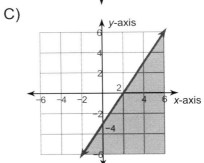

D)

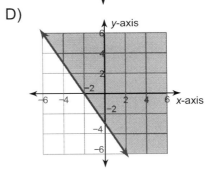

31) Which graph represents the linear inequality and its solution.

$$3x + 4y > 16$$

A)
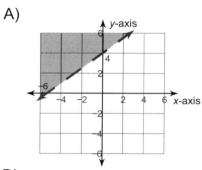

B)
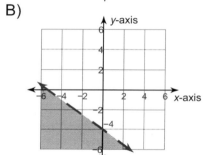

C)
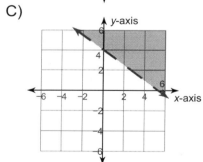

D)
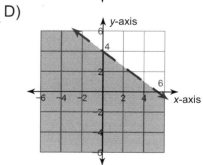

32) Which graph represents the linear inequality and its solution.

$$3x - 4y \leq 16$$

A)
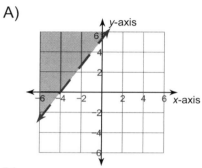

B)
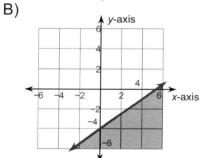

C)
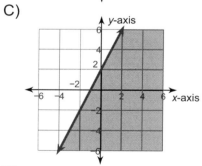

D)
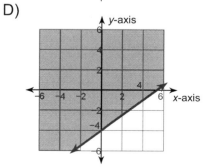

33) Which graph represents the linear inequality and its solution.

$$x \geq -3$$

A)
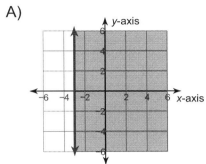

B)
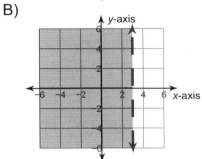

C)
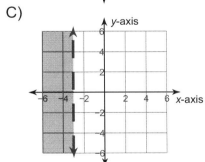

D)
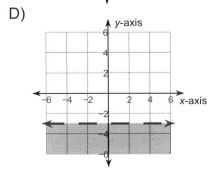

34) Which graph represents the linear inequality and its solution.

$$5x - 4y > -8$$

A)
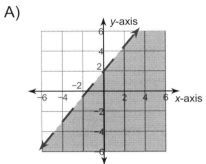

B)
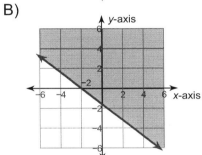

C)
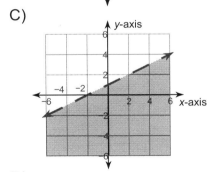

D)
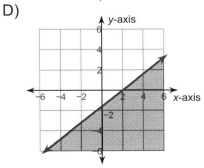

www.math-knots.com | www.a4ace.com

35) Which graph represents the linear inequality and its solution.

$$x - y > 2$$

A)
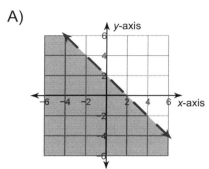

B)
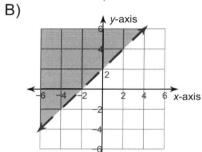

C)
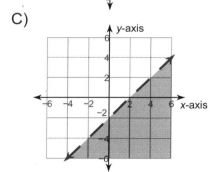

D)
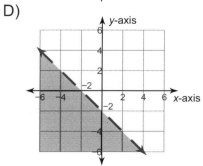

36) Which graph represents the linear inequality and its solution.

$$5x - 2y \leq -6$$

A)
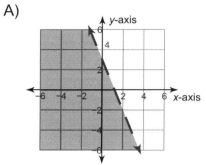

B)
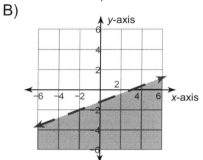

C)
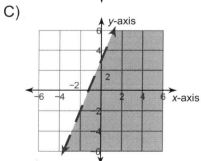

D)
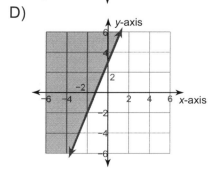

www.math-knots.com | www.a4ace.com

37) Which graph represents the linear inequality and its solution.

$$2x + 5y \leq 0$$

A)

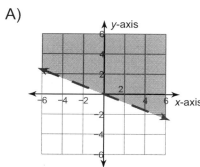

B)

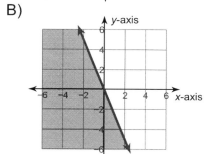

C)

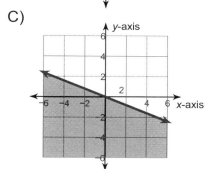

D)

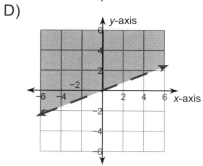

38) Which graph represents the linear inequality and its solution.

$$6x - 5y \leq -5$$

A)

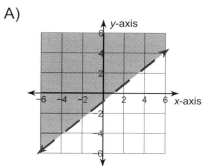

B)

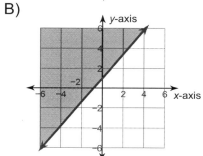

C)

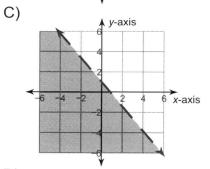

D)

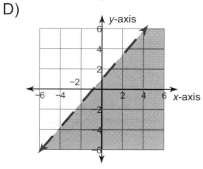

www.math-knots.com | www.a4ace.com

39) Which graph represents the linear inequality and its solution.

$$x - 2y < -8$$

A)

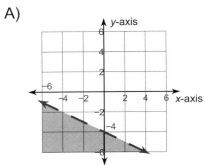

B)

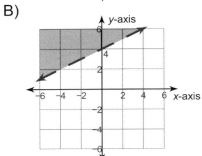

C)

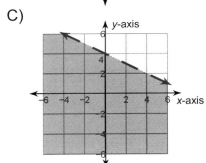

D)

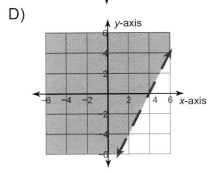

40) Which graph represents the linear inequality and its solution.

$$x \leq 3$$

A)

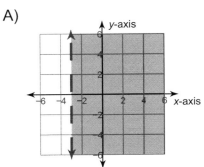

B)

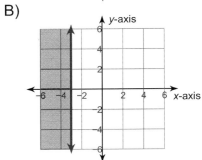

C)

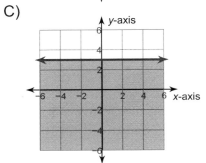

D)

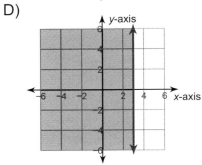

www.math-knots.com | www.a4ace.com

41) Which graph represents the linear inequality and its solution.

$$x - 3y \leq 3$$

A)
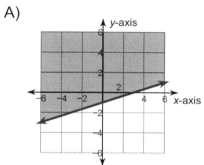

B)
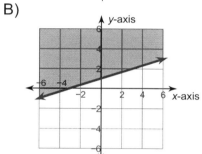

C)
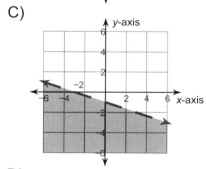

D)
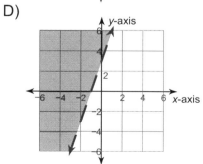

42) Which graph represents the linear inequality and its solution.

$$2x - 5y \geq 25$$

A)
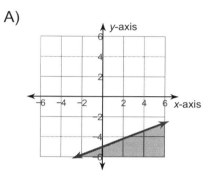

B)
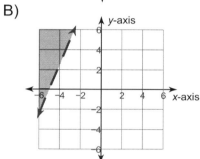

C)
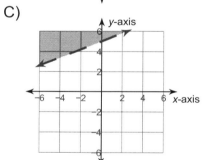

D)
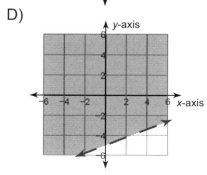

Algebra 1

43) Which graph represents the linear inequality and its solution.

$$y \geq 0$$

A)

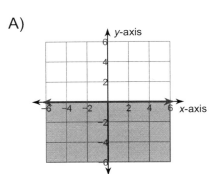

B)

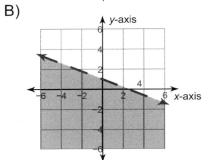

C)

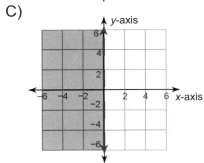

D)

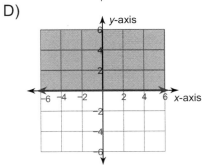

44) Which graph represents the linear inequality and its solution.

$$2x - 3y \geq 3$$

A)

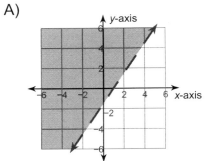

B)

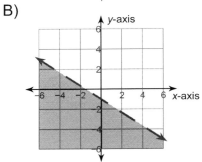

C)

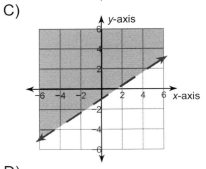

D)

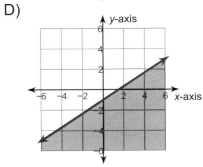

www.math-knots.com | www.a4ace.com

Algebra 1

45) Which graph represents the linear inequality and its solution.

$$5x - 2y < 0$$

A)
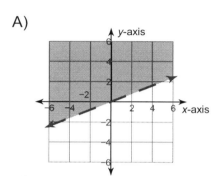

B)
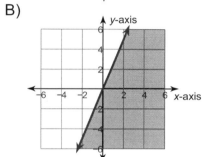

C)
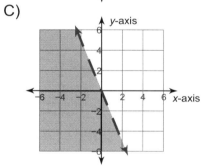

D)
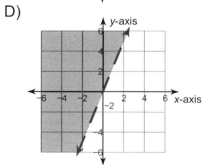

46) Which graph represents the linear inequality and its solution.

$$7x - 4y < 8$$

A)
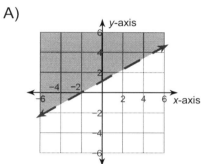

B)
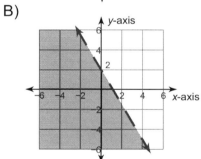

C)
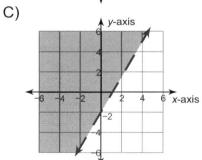

D)
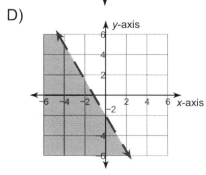

www.math-knots.com | www.a4ace.com

47) Which graph represents the linear inequality and its solution.

$$x < -4$$

A)

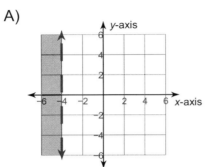

B)

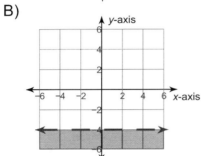

C)

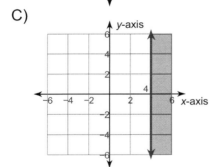

D)

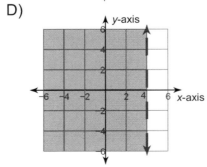

48) Which graph represents the linear inequality and its solution.

$$3x - 4y > -8$$

A)

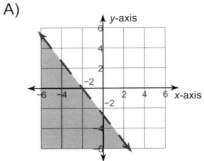

B)

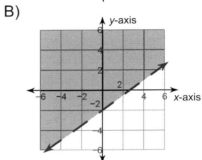

C)

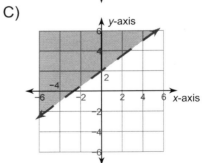

D)

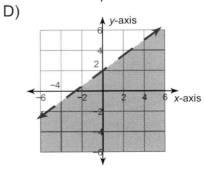

www.math-knots.com | www.a4ace.com

49) Which graph represents the linear inequality and its solution.

$$x + y \leq -2$$

A)

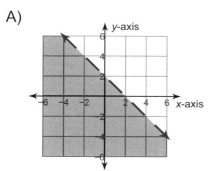

B)

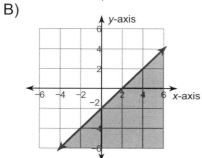

C)

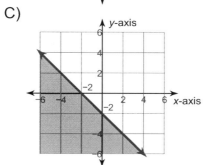

D)

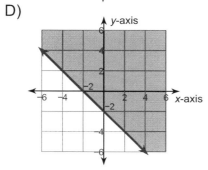

50) Which graph represents the linear inequality and its solution.

$$3x + 4y < 20$$

A)

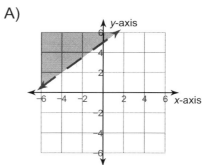

B)

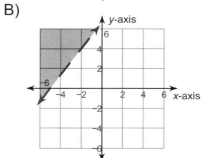

C)

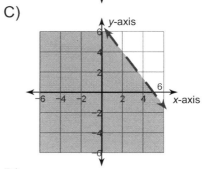

D)

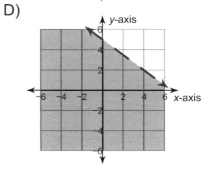

www.math-knots.com | www.a4ace.com

51) Which graph represents the linear
    inequality and its solution.

$$3x + 5y \leq 15$$

A)
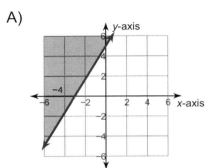

B)
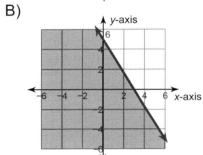

C)
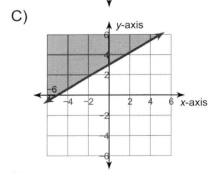

D)
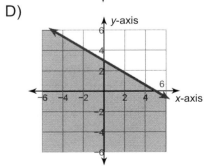

52) Which graph represents the linear
    inequality and its solution.

$$x + y < 2$$

A)
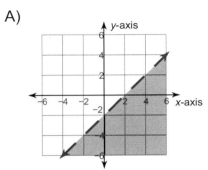

B)
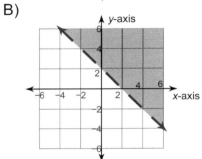

C)
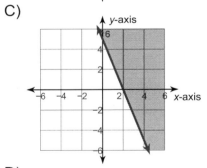

D)
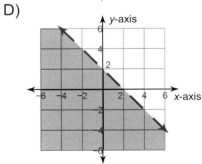

53) Which graph represents the linear inequality and its solution.

$$4x - 3y \geq 6$$

A)

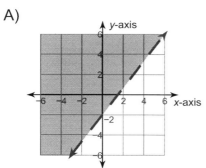

B)

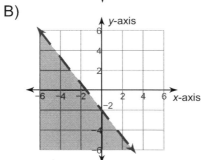

C)

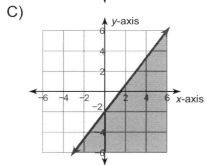

D)

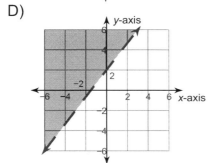

54) Which graph represents the linear inequality and its solution.

$$4x - y \geq 2$$

A)

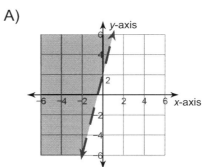

B)

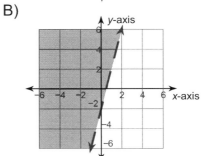

C)

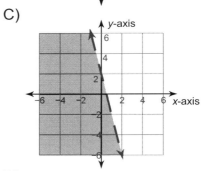

D)

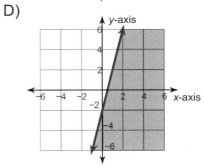

55) Which graph represents the linear inequality and its solution.

$$x - 4y \geq -4$$

A)

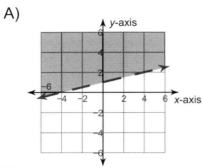

B)

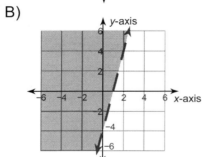

C)

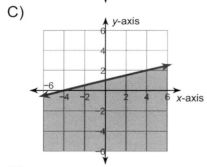

D)

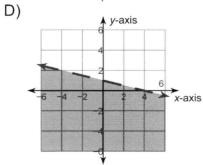

56) Which graph represents the linear inequality and its solution.

$$2x - y < 1$$

A)

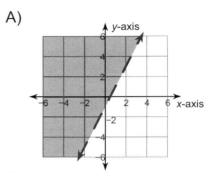

B)

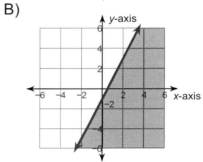

C)

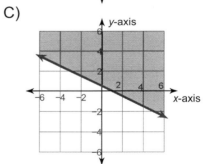

D)

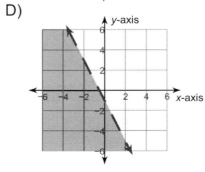

www.math-knots.com | www.a4ace.com

57) Which graph represents the linear inequality and its solution.

$$6x - 5y \geq -5$$

A)

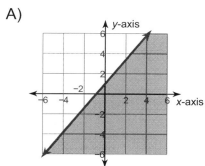

B)

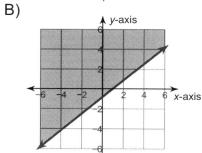

C)

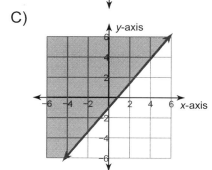

D)

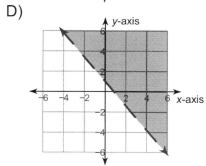

58) Which graph represents the linear inequality and its solution.

$$2x + y > 3$$

A)

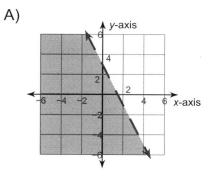

B)

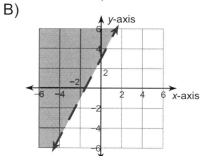

C)

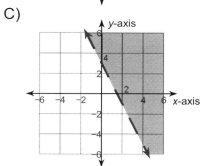

D)

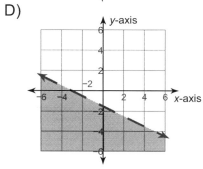

www.math-knots.com | www.a4ace.com

59) Which graph represents the linear inequality and its solution.

$$x - 4y \geq 0$$

A)
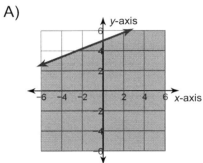

B)
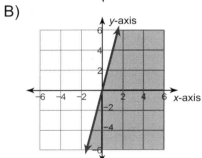

C)
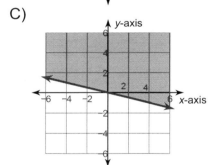

D)
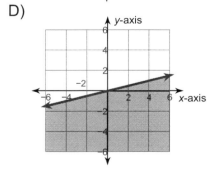

60) Which graph represents the linear inequality and its solution.

$$9x + 5y \geq 25$$

A)
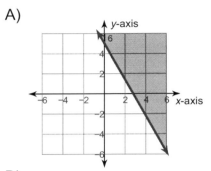

B)
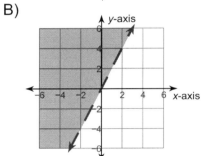

C)
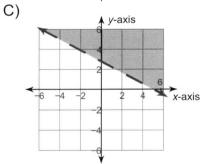

D)
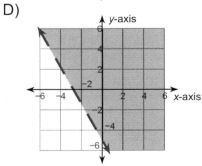

www.math-knots.com | www.a4ace.com

61) Which graph represents the linear inequality and its solution.

$$x \le -2$$

A)
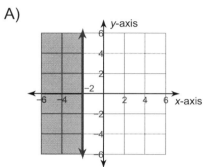

B)
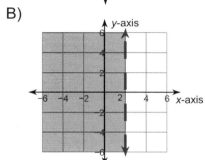

C)
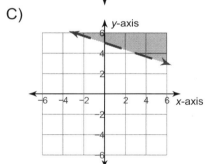

D)
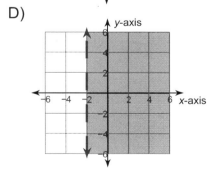

62) Which graph represents the linear inequality and its solution.

$$3x + y > 3$$

A)
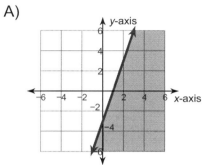

B)
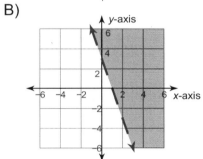

C)
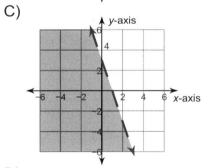

D)
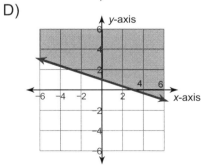

www.math-knots.com | www.a4ace.com

63) Which graph represents the linear
inequality and its solution.

$$x < 1$$

A)

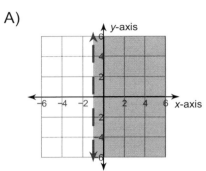

B)

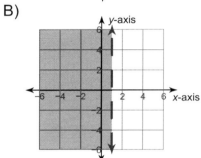

C)

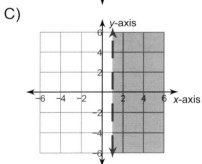

D)

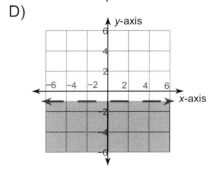

64) Which graph represents the linear
inequality and its solution.

$$x - y \geq 1$$

A)

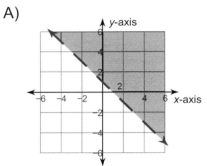

B)

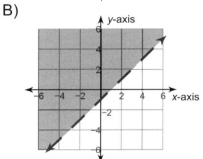

C)

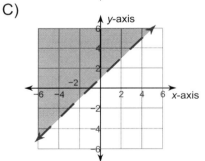

D)

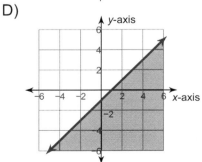

www.math-knots.com | www.a4ace.com

65) Solve the inequality.

$$|n + 7| \geq 15$$

66) Solve the inequality.

$$|x + 8| \geq 1$$

67) Solve the inequality.

$$\left|\frac{n}{3}\right| < 2$$

68) Solve the inequality.

$$\left|\frac{k}{8}\right| \leq 5$$

69) Solve the inequality.

$$\left|\frac{x}{10}\right| < 2$$

70) Solve the inequality.

$$|x - 7| > 6$$

71) Solve the inequality.

$$|6 + p| \geq 7$$

72) Solve the inequality.

$$\left|\frac{x}{9}\right| > 2$$

73) Solve the inequality.

$$\left|\frac{x}{8}\right| \geq 2$$

74) Solve the inequality.

$$|8 + x| \leq 15$$

75) Solve the inequality.

$$|-4p| \geq 12$$

76) Solve the inequality.

$$|k + 3| \geq 11$$

77) Solve the inequality.

$$|8v| \leq 48$$

78) Solve the inequality.

$$|n + 9| \geq 12$$

79) Solve the inequality.

$$|-3a| > 24$$

80) Solve the inequality.

$$|7a| > 21$$

81) Solve the inequality.

$$|x - 2| > 4$$

82) Solve the inequality.

$$|v - 3| < 6$$

83) Solve the inequality.

$$|3x| \leq 27$$

84) Solve the inequality.

$$\left|\frac{a}{9}\right| < 1$$

www.math-knots.com | www.a4ace.com

85) Solve the inequality.

$$|9 + n| < 1$$

86) Solve the inequality.

$$|10b| + 4 > 94$$

87) Solve the inequality.

$$|7 + r| + 7 < 15$$

88) Solve the inequality.

$$|-10a| - 9 > 91$$

89) Solve the inequality.

$$|x - 9| + 9 \geq 13$$

90) Solve the inequality.

$$3|-1 + x| > 12$$

91) Solve the inequality.

$$-6 + |r + 1| \geq 1$$

92) Solve the inequality.

$$\frac{|-9r|}{4} < 3$$

93) Solve the inequality.

$$\frac{|10x|}{7} \geq 2$$

94) Solve the inequality.

$$|x + 10| - 1 < 10$$

95) Solve the inequality.

$$10 + |x - 4| > 11$$

96) Solve the inequality.

$$\left| \frac{n}{7} \right| - 6 \geq -5$$

97) Solve the inequality.

$$2 + \left| \frac{m}{8} \right| \leq 3$$

98) Solve the inequality.

$$2 + \left| \frac{k}{6} \right| < 3$$

99) Solve the inequality.

$$\frac{|-1 + x|}{3} < 3$$

100) Solve the inequality.

$$|x - 5| - 8 \leq -4$$

101) Solve the inequality.

$$-2 + \left| \frac{v}{7} \right| > -1$$

102) Solve the inequality.

$$|x - 6| + 10 > 19$$

103) Solve the inequality.

$$|x + 4| - 1 < 6$$

104) Solve the inequality.

$$-10 \left| \frac{n}{4} \right| \leq -5$$

105) Solve the inequality.

$$4\left|p+9\right|<72$$

106) Solve the inequality.

$$10+\left|-8+m\right|\le 15$$

107) Solve the inequality.

$$4+9\left|-3x\right|\ge 112$$

108) Solve the inequality.

$$\left|5p\right|+10>20$$

109) Solve the inequality.

$$-9\left|4+k\right|-9>-18$$

110) Solve the inequality.

$$-8\left|5+x\right|+5>-107$$

111) Solve the inequality.

$$3-4\left|10k\right|>-117$$

112) Solve the inequality.

$$8\left|1+x\right|+4\ge 60$$

113) Solve the inequality.

$$8\left|\frac{k}{2}\right|+6\le 10$$

114) Solve the inequality.

$$5\left|\frac{x}{6}\right|-6\ge -1$$

115) Solve the inequality.

$$7\left|-4 + x\right| + 9 \le 86$$

116) Solve the inequality.

$$7 - \left|-3x\right| > -8$$

117) Solve the inequality.

$$6 - 7\left|\frac{v}{10}\right| \le -1$$

118) Solve the inequality.

$$-9\left|-2 + a\right| + 4 \le -41$$

119) Solve the inequality.

$$7\left|\frac{v}{9}\right| + 8 < 15$$

120) Solve the inequality.

$$4\left|\frac{x}{10}\right| - 4 < -2$$

121) Solve the inequality.

$$-8 + 8\left|-8 + n\right| \le 80$$

122) Solve the inequality.

$$-3 + 6\left|3m\right| > 87$$

123) Solve the inequality.

$$9\left|-8 + x\right| + 6 > 78$$

124) Solve the inequality.

$$9 + 2\left|6n\right| \ge 117$$

125) Solve the inequality.

$$10\left|2x\right| - 10 \geq 90$$

126) Solve the inequality.

$$2\left|2 + k\right| + 8 > 22$$

127) Solve the inequality.

$$6\left|\dfrac{x}{6}\right| - 5 \leq -3$$

128) Solve the inequality.

$$5\left|n - 2\right| + 6 \geq 36$$

129) Solve the inequality.

$$2\left|\dfrac{v}{8}\right| - 5 < -4$$

130) Solve the inequality.

$$3 + 9\left|-3b\right| \leq 84$$

131) Solve the inequality.

$$-3\left|10b\right| + 1 \leq -59$$

132) Solve the inequality.

$$\left|-5p + 9\right| < 26$$

133) Solve the inequality.

$$\left|-9k + 4\right| \leq 76$$

134) Solve the inequality.

$$\left|9 - 3n\right| \geq 15$$

135) Solve the inequality.

$$\left|-2n + 1\right| \le 5$$

136) Solve the inequality.

$$\left|-6n + 2\right| > 14$$

137) Solve the inequality.

$$\left|9 - 3n\right| \le 21$$

138) Solve the inequality.

$$\left|9k + 2\right| \le 29$$

139) Solve the inequality.

$$\left|10a + 5\right| < 35$$

140) Solve the inequality.

$$\left|-5 + 2b\right| \le 15$$

141) Solve the inequality.

$$\left|9 + a\right| \ge 10$$

142) Solve the inequality.

$$\left|8n + 5\right| \ge 37$$

143) Solve the inequality.

$$\left|8n - 8\right| \ge 32$$

144) Solve the inequality.

$$\left|-4m + 1\right| < 15$$

www.math-knots.com | www.a4ace.com

145) Solve the inequality.

$$|2 - 3r| \leq 10$$

146) Solve the inequality.

$$|n - 2| \geq 2$$

147) Solve the inequality.

$$|6 + 2x| \leq 26$$

148) Solve the inequality.

$$|10 + 2n| > 6$$

149) Solve the inequality.

$$|10 - 8n| \geq 34$$

150) Solve the inequality.

$$|4m - 4| \leq 12$$

151) Solve the inequality.

$$|9x + 10| \leq 44$$

152) Solve the inequality.

$$|10n - 4| < 104$$

153) Solve the inequality.

$$|10 - 4b| > 6$$

154) Solve the inequality.

$$|4 - 9n| \leq 67$$

155) Solve the inequality.

$$\left|-9-6v\right| \le 69$$

156) Solve the inequality.

$$\left|8a-8\right| > 16$$

157) Solve the inequality.

$$\frac{\left|7a+6\right|}{6} \ge 2$$

158) Solve the inequality.

$$4 + \left|1-k\right| < 8$$

159) Solve the inequality.

$$\left|7x+6\right| + 9 < 45$$

160) Solve the inequality.

$$6\left|2+6a\right| > 48$$

161) Solve the inequality.

$$\left|-2+6n\right| - 4 \ge 24$$

162) Solve the inequality.

$$4 + \left|-8x-5\right| > 7$$

163) Solve the inequality.

$$\left|10-9a\right| - 10 \le 61$$

164) Solve the inequality.

$$7\left|6+7p\right| \ge 7$$

165) Solve the inequality.

$$|7 - 7n| + 5 > 33$$

170) Solve the inequality.

$$|3x + 6| - 9 \le 0$$

166) Solve the inequality.

$$2|b + 6| > 2$$

171) Solve the inequality.

$$7|9 - 6k| > 21$$

167) Solve the inequality.

$$\frac{|7x + 5|}{2} \ge 3$$

172) Solve the inequality.

$$|-1 - 7m| - 7 < 1$$

168) Solve the inequality.

$$|-n - 7| + 9 \ge 10$$

173) Solve the inequality.

$$3 + |4k - 9| \le 10$$

169) Solve the inequality.

$$3 + |9 - 2v| \le 24$$

174) Solve the inequality.

$$9 + |6 + 8x| \le 67$$

175) Solve the inequality.

$$\frac{|2 + 2n|}{7} \leq 5$$

176) Solve the inequality.

$$8 + |-2 - 9x| \leq 55$$

177) Solve the inequality.

$$\frac{|-10n - 1|}{7} < 1$$

178) Solve the inequality.

$$\frac{|6 + 4k|}{2} \leq 4$$

179) Solve the inequality.

$$\frac{|7k - 5|}{5} \leq 5$$

180) Solve the inequality.

$$6 + 6|2 + p| \leq 36$$

181) Solve the inequality.

$$-5|3 + 9x| + 5 > -25$$

182) Solve the inequality.

$$4|-x - 9| - 5 > 31$$

183) Solve the inequality.

$$-2 - 2|9x + 2| < -88$$

184) Solve the inequality.

$$|8 + 9x| + 3 \leq 13$$

 **Algebra 1**

185) Solve the inequality.

$$5 - 9\left|-6a + 3\right| > -76$$

186) Solve the inequality.

$$-4\left|8 - 8n\right| + 1 > -95$$

187) Solve the inequality.

$$5 + \left|5b + 8\right| > 18$$

188) Solve the inequality.

$$5\left|2 + 9b\right| - 3 \geq 77$$

189) Solve the inequality.

$$8 - 5\left|5 + 7b\right| \geq -87$$

190) Solve the inequality.

$$3 - \left|5n + 4\right| \leq -28$$

191) Solve the inequality.

$$-\left|7 + 6n\right| + 3 < -20$$

192) Solve the inequality.

$$8 - 8\left|7a + 6\right| < -112$$

193) Solve the inequality.

$$-7\left|9x - 4\right| + 7 > -28$$

194) Solve the inequality.

$$4\left|4 + 3p\right| - 6 < 10$$

195) Solve the inequality.

$$3 - 6|n + 8| \leq -81$$

196) Solve the inequality.

$$|8n - 2| - 4 < 58$$

197) Solve the inequality.

$$7 + 3|1 + 2n| < 10$$

www.math-knots.com | www.a4ace.com

1) Evaluate the below function.

$$f(a) = 4a - 4; \text{ Find } f(9)$$

A) 12   B) 0

C) 32   D) 8

2) Evaluate the below function.

$$f(x) = 4x + 3; \text{ Find } f(-1)$$

A) −1   B) −25

C) 43   D) 3

3) Evaluate the below function.

$$g(x) = x + 3; \text{ Find } g(10)$$

A) 2   B) 1

C) 13   D) 12

4) Evaluate the below function.

$$p(n) = 4n + 3; \text{ Find } p(-6)$$

A) −21   B) −5

C) 19   D) 39

5) Evaluate the below function.

$$k(a) = 4a - 2; \text{ Find } k(1)$$

A) 38   B) −34

C) 2   D) −6

6) Evaluate the below function.

$$k(x) = 4x - 3; \text{ Find } k(-1)$$

A) −39   B) 5

C) −43   D) −7

7) Evaluate the below function.

$$g(n) = 3n + 3; \text{ Find } g(10)$$

A) 30   B) −27

C) 33   D) 27

8) Evaluate the below function.

$$k(n) = 4n + 2; \text{ Find } k(-7)$$

A) −34   B) −6

C) −26   D) 6

9) Evaluate the below function.

$$k(x) = 2x^3 + x^2; \text{ Find } k(4)$$

A) −1   B) −12

C) 20   D) 144

10) Evaluate the below function.

$$f(a) = a^2 + 1; \text{ Find } f(-10)$$

A) 10   B) 101

C) 17   D) 2

11) Evaluate the below function.

$$w(t) = t^2 - 3; \text{Find } w(-4)$$

A) 13   B) 1

C) 46   D) 78

12) Evaluate the below function.

$$h(a) = a^3 - a^2; \text{Find } h(-3)$$

A) −36   B) 18

C) −80   D) −150

13) Evaluate the below function.

$$k(t) = -t^2 + t; \text{Find } k(-2)$$

A) −12   B) −6

C) −42   D) −30

14) Evaluate the below function.

$$f(t) = t^2 + 1 + t; \text{Find } f(3)$$

A) 111   B) 13

C) 57   D) 1

15) Evaluate the below function.

$$f(n) = n^2 - 2; \text{Find } f(-6)$$

A) 34   B) 2

C) 62   D) 79

16) Evaluate the below function.

$$p(n) = -3n^3 - 5; \text{Find } p(-4)$$

A) 76   B) −29

C) −8   D) 187

17) Evaluate the below function.

$$k(x) = 3x + 2; \text{Find } k(5)$$

A) 17   B) 26

C) 11   D) −19

18) Evaluate the below function.

$$h(n) = n^2 + 3n; \text{Find } h(8)$$

A) 10   B) 4

C) 40   D) 88

19) Evaluate the below function.

$$f(x) = 3x + 1; \text{Find } f(8)$$

A) 10   B) 25

C) −29   D) −14

20) Evaluate the below function.

$$h(n) = 2n - 4; \text{Find } h(10)$$

A) 16   B) 8

C) −24   D) 12

21) Evaluate the below function.

$$f(n) = n^2 - n; \text{ Find } f(-4)$$

A) 20   B) 110

C) 90   D) 42

22) Evaluate the below function.

$$p(n) = -2n^2 - 3; \text{ Find } p(5)$$

A) –131   B) –165

C) –75   D) –53

23) Evaluate the below function.

$$h(n) = |n|; \text{ Find } h(8)$$

A) 3   B) 6

C) 2   D) 8

24) Evaluate the below function.

$$p(t) = |-3t - 1|; \text{ Find } p(-6)$$

A) 23   B) 7

C) 22   D) 17

25) Evaluate the below function.

$$h(x) = |-3x|; \text{ Find } h(-10)$$

A) 30   B) 9

C) 18   D) 24

26) Evaluate the below function.

$$k(t) = -2|t| - 1; \text{ Find } k(-8)$$

A) −21   B) −1

C) −17   D) −19

27) Evaluate the below function.

$$f(a) = 3|2a|; \text{ Find } f(4)$$

A) 54   B) 24

C) 12   D) 42

28) Evaluate the below function.

$$h(x) = |2x + 2|; \text{ Find } h(4)$$

A) 18   B) 10

C) 12   D) 4

29) Evaluate the below function.

$$h(x) = 3|x|; \text{ Find } h(2)$$

A) 24   B) 0

C) 3   D) 6

30) Evaluate the below function.

$$w(n) = |n + 2| - 2; \text{ Find } w(9)$$

A) 6   B) 9

C) −2   D) 7

31) Evaluate the below function.

$f (a) = | a - 3 |$; Find $f (- 3)$

A) 13   B) 6

C) 3   D) 4

32) Evaluate the below function.

$g (x) = | x |$; Find $g (7)$

A) 7   B) -2

C) -10   D) -9

33) Evaluate the below function.

$g (x) = | x | - 2$; Find $g (9)$

A) 2   B) 7

C) 4   D) -1

34) Evaluate the below function.

$k (n) = - | n + 3 |$; Find $k (3)$

A) -12   B) -6

C) -5   D) -7

35) Evaluate the below function.

$w (n) = | n | + 3$; Find $w (- 10)$

A) 6   B) 12

C) 9   D) 13

36) Evaluate the below function.

$h (n) = | n | - 3$; Find $h (10)$

A) 7   B) 2

C) 4   D) -1

37) Evaluate the below function.

$h (x) = | 2 x |$; Find $h (- 6)$

A) 12   B) 18

C) 8   D) 10

38) Evaluate the below function.

$h (n) = - 2 | 2 n |$; Find $h (5)$

A) -20   B) -16

C) -32   D) -12

39) Evaluate the below function.

$f (a) = | - 3 a + 1 | - 2$; Find $f (5)$

A) 24   B) 5

C) 12   D) 17

40) Evaluate the below function.

$g (a) = - 2 | 3 a | + 2$; Find $g (- 6)$

A) -46   B) -10

C) -34   D) -16

41) Evaluate the below function.

$k(n) = |n - 2| - 2$; Find $k(-2)$

A) 6   B) 2

C) 4   D) 1

42) Evaluate the below function.

$h(n) = |n|$; Find $h(1)$

A) 10   B) 4

C) 3   D) 1

43) Evaluate the below function.

$h(t) = |-3t + 3| - 2$; Find $h(-6)$

A) 16   B) 19

C) -2   D) 25

44) Evaluate the below function.

$g(a) = -2|a|$; Find $g(4)$

A) -6   B) -18

C) -8   D) -10

45) Evaluate the below function.

$w(n) = 5^n$; Find $w(0)$

A) 1   B) $\dfrac{1}{5}$

C) $\dfrac{1}{25}$ D) 5

46) Evaluate the below function.

$g(t) = 2^t - 1$; Find $g(-2)$

A) $-\dfrac{3}{4}$ B) 1

C) 3   D) $-\dfrac{1}{2}$

47) Evaluate the below function.

$k(a) = 3 \cdot 5^{3a} + 2$; Find $k(0)$

A) 46877   B) $\dfrac{253}{125}$

C) 5   D) $\dfrac{31253}{15625}$

48) Evaluate the below function.

$h(x) = 4^x - 3$; Find $h(2)$

A) 1   B) 13

C) $-\dfrac{47}{16}$ D) $-\dfrac{11}{4}$

49) Evaluate the below function.

$f(n) = 2^{n+1}$; Find $f(0)$

A) 2   B) $\dfrac{1}{2}$

C) 1   D) 8

50) Evaluate the below function.

$g(n) = 2 \cdot 3^{-n-1} + 3$; Find $g(-2)$

A) $\dfrac{29}{9}$ B) $\dfrac{11}{3}$

C) $\dfrac{83}{27}$ D) 9

51) Evaluate the below function.

$f(n) = -2 \cdot 2^{-n+3}$; Find $f(1)$

A) −4 B) −32

C) −16 D) −8

52) Evaluate the below function.

$p(x) = 4^x + 1$; Find $p(1)$

A) $\dfrac{5}{4}$ B) $\dfrac{17}{16}$

C) 5 D) 2

53) Evaluate the below function.

$w(x) = -3 \cdot 3^x + 3$; Find $w(1)$

A) 2 B) −6

C) $\dfrac{8}{3}$ D) −24

54) Evaluate the below function.

$w(x) = 2^{-x}$; Find $w(2)$

A) 4 B) 1

C) $\dfrac{1}{4}$ D) 2

55) Evaluate the below function.

$g(x) = 5^{x+2}$; Find $g(-2)$

A) 5 B) 125

C) 25 D) 1

56) Evaluate the below function.

$f(x) = 4^{x-1}$; Find $f(0)$

A) 4 B) $\dfrac{1}{16}$

C) $\dfrac{1}{4}$ D) 1

57) Evaluate the below function.

$h(n) = -3^n$; Find $h(0)$

A) −1 B) −9

C) $-\dfrac{1}{3}$ D) −3

58) Evaluate the below function.

$f(t) = -2^{t+2} - 1$; Find $f(-2)$

A) −2 B) −9

C) −5 D) −3

59) Evaluate the below function.

$f(x) = -2 \cdot 3^{x+1}$; Find $f(0)$

A) $-\dfrac{2}{3}$ B) −18

C) −2 D) −6

60) Evaluate the below function.

$g(x) = -3 \cdot 4^{2x+1}$; Find $g(1)$

A) −192 B) −12

C) $-\dfrac{3}{4}$ D) $-\dfrac{3}{64}$

61) Evaluate the below function.

$h(x) = 4^x$; Find $h(-1)$

A) 16
B) $\dfrac{1}{4}$

C) 4
D) $\dfrac{1}{16}$

62) Evaluate the below function.

$w(x) = 4^{x-1} - 1$; Find $w(-1)$

A) $-\dfrac{3}{4}$
B) $-\dfrac{15}{16}$

C) 3
D) $-\dfrac{63}{64}$

63) Evaluate the below function.

$f(x) = -3^{x+1}$; Find $f(-2)$

A) −9
B) $-\dfrac{1}{3}$

C) −1
D) −27

64) Evaluate the below function.

$f(t) = 4^{t+3} + 1$; Find $f(-2)$

A) 17
B) 1025

C) 5
D) 65

65) Evaluate the below function.

$g(a) = 3 \cdot 2^{-a} - 2$; Find $g(-2)$

A) 4
B) $-\dfrac{1}{2}$

C) $-\dfrac{5}{4}$
D) 10

66) Evaluate the below function.

$f(a) = 2^{3a} - 2$; Find $f(0)$

A) 62
B) $-\dfrac{127}{64}$

C) −1
D) 6

67) Evaluate the below function.

$h(n) = 5^n$; Find $h(-1)$

A) 5
B) 1

C) $\dfrac{1}{5}$
D) 25

68) Evaluate the below function.

$w(t) = \dfrac{1}{3}t + \dfrac{1}{2}$; Find $w\left(\dfrac{1}{5}\right)$

A) $\dfrac{5}{6}$
B) $\dfrac{17}{30}$

C) $\dfrac{17}{18}$
D) $\dfrac{29}{30}$

69) Evaluate the below function.

$f(x) = |x|$; Find $f\left(\dfrac{3}{2}\right)$

A) $\dfrac{4}{5}$
B) $\dfrac{15}{8}$

C) $\dfrac{3}{2}$
D) $\dfrac{7}{6}$

70) Evaluate the below function.

$f(x) = 3^{x+1}$; Find $f(0)$

A) 3
B) 9

C) 27
D) 1

71) Evaluate the below function.

$$k(n) = \frac{1}{2}n^2 + \frac{4}{5}; \text{ Find } k\left(\frac{1}{2}\right)$$

A) $\frac{77}{40}$ B) $\frac{4}{5}$

C) $\frac{13}{10}$ D) $\frac{37}{40}$

72) Evaluate the below function.

$$g(x) = x^2 - \frac{1}{3}; \text{ Find } g(-2)$$

A) $\frac{11}{3}$ B) $\frac{13}{9}$

C) $\frac{1}{9}$ D) $-\frac{1}{3}$

73) Evaluate the below function.

$$h(n) = n + \frac{1}{2}; \text{ Find } h\left(\frac{6}{5}\right)$$

A) $\frac{19}{10}$ B) 1

C) $-\frac{1}{6}$ D) $\frac{17}{10}$

74) Evaluate the below function.

$$h(a) = \left|a - \frac{4}{3}\right|; \text{ Find } h\left(\frac{7}{5}\right)$$

A) $\frac{8}{3}$ B) $\frac{1}{3}$

C) $\frac{67}{21}$ D) $\frac{1}{15}$

75) Evaluate the below function.

$$g(x) = x^2 - \frac{1}{3}x; \text{ Find } g(1)$$

A) $\frac{7}{4}$ B) 34

C) 8   D) $\frac{2}{3}$

76) Evaluate the below function.

$$g(n) = \frac{1}{3}n - \frac{5}{4}; \text{ Find } g(-2)$$

A) $-\frac{53}{36}$ B) $-\frac{23}{12}$

C) $-\frac{43}{60}$ D) $-\frac{13}{20}$

77) Evaluate the below function.

$$f(a) = 1 + \frac{1}{2}a; \text{ Find } f\left(-\frac{4}{5}\right)$$

A) $\frac{3}{2}$ B) $\frac{3}{5}$

C) $\frac{2}{9}$ D) $\frac{3}{4}$

78) Evaluate the below function.

$$f(t) = 2t - \frac{5}{3}; \text{ Find } f(2)$$

A) -1 B) $-\frac{2}{3}$

C) $\frac{7}{3}$ D) $\frac{1}{3}$

79) Evaluate the below function.

$$w(n) = 3^n + \frac{2}{3}; \text{ Find } w(-1)$$

A) $\frac{29}{3}$ B) $\frac{11}{3}$

C) 1   D) $\frac{5}{3}$

80) Evaluate the below function.

$$w(t) = \left|2t - \frac{5}{3}\right|; \text{ Find } w\left(\frac{9}{10}\right)$$

A) $\frac{49}{3}$ B) $\frac{5}{3}$

C) $\frac{2}{15}$ D) 5

Algebra 1

81) Evaluate the below function.

$$h(a) = 3 + \frac{3}{4}a; \text{ Find } h(-2)$$

A) $\frac{27}{10}$ B) $\frac{13}{6}$

C) $\frac{15}{7}$ D) $\frac{3}{2}$

82) Evaluate the below function.

$$k(t) = -\left|\frac{2}{3}t\right|; \text{ Find } k(1)$$

A) $-\frac{2}{3}$ B) $-\frac{4}{3}$

C) $-\frac{7}{6}$ D) $-\frac{16}{15}$

83) Evaluate the below function.

$$g(x) = -\frac{1}{3} \cdot \left|x - \frac{5}{3}\right| - 1; \text{ Find } g(-1)$$

A) $-\frac{10}{9}$ B) $-\frac{73}{45}$

C) $-\frac{17}{9}$ D) $-\frac{128}{63}$

84) Evaluate the below function.

$$h(n) = \left|n - \frac{1}{2}\right| + 1; \text{ Find } h(1)$$

A) $\frac{17}{8}$ B) $\frac{3}{2}$

C) $\frac{5}{2}$ D) $\frac{7}{3}$

85) Evaluate the below function.

$$f(t) = \frac{1}{3}t; \text{ Find } f\left(\frac{1}{10}\right)$$

A) $\frac{4}{9}$ B) $-\frac{1}{3}$

C) $\frac{1}{30}$ D) $\frac{5}{12}$

86) Evaluate the below function.

$$f(x) = 4^x + \frac{5}{3}; \text{ Find } f(0)$$

A) $\frac{83}{48}$ B) $\frac{53}{3}$

C) $\frac{8}{3}$ D) $\frac{17}{3}$

87) Evaluate the below function.

$$k(x) = \left|\frac{1}{2}x + \frac{3}{2}\right|; \text{ Find } k\left(-\frac{4}{3}\right)$$

A) $\frac{5}{4}$ B) 2

C) $\frac{5}{6}$ D) $\frac{5}{2}$

88) Evaluate the below function.

$$h(x) = 2^{x-2}; \text{ Find } h(-1)$$

A) $\frac{1}{4}$ B) 1

C) $\frac{1}{8}$ D) $\frac{1}{2}$

89) Evaluate the below function.

$$g(x) = x^3 + 3; \text{ Find } g\left(-\frac{3}{4}\right)$$

A) 4 B) $-\frac{44}{27}$

C) $\frac{165}{64}$ D) $\frac{193}{64}$

90) Evaluate the below function.

$$f(t) = -t - \frac{2}{5}; \text{ Find } f\left(-\frac{12}{7}\right)$$

A) $\frac{67}{45}$ B) $\frac{3}{5}$

C) $\frac{46}{35}$ D) $\frac{19}{15}$

91) Evaluate the below function.

$$h(x) = x^2 - 2 - 2x; \text{ Find } h(-2)$$

A) $-\dfrac{11}{4}$　　　B) -3

C) 6　　　D) $-\dfrac{59}{36}$

92) Evaluate the below function.

$$g(a) = |-a - 1|; \text{ Find } g\left(-\dfrac{1}{2}\right)$$

A) 1　　　B) $\dfrac{5}{3}$

C) $\dfrac{7}{10}$　　　D) $\dfrac{1}{2}$

93) Evaluate the below function.

$$w(x) = x^3 - 4x; \text{ Find } w(-2x)$$

A) $-x + \dfrac{1}{64}x^3$　　　B) $-8x^3 + 8x$

C) $64x^3 - 16x$　　　D) $-27x^3 + 12x$

94) Evaluate the below function.

$$g(x) = 2x - 4; \text{ Find } g(4x)$$

A) $-4x - 4$　　　B) $-2 - 2x$

C) $-4 + \dfrac{1}{2}x$　　　D) $8x - 4$

95) Evaluate the below function.

$$f(t) = -t^2 - 7t; \text{ Find } f\left(\dfrac{t}{4}\right)$$

A) $6 + 5t - t^2$　　　B) $-t^4 - 7t^2$

C) $-t^2 + 7t$　　　D) $-\dfrac{1}{16}t^2 - \dfrac{7}{4}t$

96) Evaluate the below function.

$$k(x) = 3x - 4; \text{ Find } k(-4x)$$

A) $3x + 5$　　　B) $-3x - 4$

C) $-12x - 4$　　　D) $3x - 13$

97) Evaluate the below function.

$$k(n) = n^3 - 2n; \text{ Find } k(n + 2)$$

A) $n^3 + 6n^2 + 10n + 4$

B) $-8n^3 + 4n$

C) $n^6 - 2n^2$

D) $n^3 - 6n^2 + 10n - 4$

98) Evaluate the below function.

$$h(a) = a^3 - 1; \text{ Find } h(-2a)$$

A) $8a^3 - 1$　　　B) $-8a^3 - 1$

C) $64a^3 - 1$　　　D) $-1 + \dfrac{1}{27}a^3$

99) Evaluate the below function.

$$f(n) = 3n - 2; \text{ Find } f\left(\dfrac{n}{4}\right)$$

A) $-5 - 3n$　　　B) $-2 + \dfrac{3}{4}n$

C) $3n - 5$　　　D) $10 - 3n$

100) Evaluate the below function.

$$p(n) = n^3 - 3; \text{ Find } p\left(\dfrac{n}{2}\right)$$

A) $n^6 - 3$　　　B) $27n^3 - 3$

C) $64n^3 - 3$　　　D) $-3 + \dfrac{1}{8}n^3$

101) Evaluate the below function.

$$w(t) = 4t + 1; \text{ Find } w\left(\frac{t}{3}\right)$$

A) $-16t + 1$ B) $1 + \frac{4}{3}t$

C) $4t - 7$ D) $-8t + 1$

102) Evaluate the below function.

$$f(n) = -3n + 3 \text{ ; Find } f(n^2)$$

A) $3 - \frac{3}{2}n$ B) $-9n + 3$

C) $-3n^2 + 3$ D) $12n + 3$

103) Evaluate the below function.

$$h(t) = t^2 + 3; \text{ Find } h(4t)$$

A) $16t^2 + 3$ B) $t^2 - 8t + 19$

C) $t^2 + 8t + 19$ D) $t^4 + 3$

104) Evaluate the below function.

$$g(t) = t^2 + 3; \text{ Find } g(-3t)$$

A) $t^4 + 3$ B) $9t^2 + 3$

C) $3 + \frac{1}{16}t^2$ D) $t^2 + 4t + 7$

105) Evaluate the below function.

$$h(t) = -4t - 5; \text{ Find } h(t^2)$$

A) $8t - 5$ B) $-4t^2 - 5$

C) $-4t - 1$ D) $-4t - 21$

106) Evaluate the below function.

$$h(x) = 2x + 4; \text{ Find } h\left(\frac{x}{2}\right)$$

A) $x + 4$ B) $-6x + 4$

C) $6x + 4$ D) $8 + 2x$

107) Evaluate the below function.

$$g(x) = -3x^3 + 5x; \text{ Find } g(-2x)$$

A) $-24x^3 + 10x$

B) $-3x^3 + 36x^2 - 139x + 172$

C) $-3x^6 + 5x^2$

D) $24x^3 - 10x$

108) Evaluate the below function.

$$p(x) = x^2 - 4; \text{ Find } p(x - 4)$$

A) $x^2 - 8x + 12$

B) $4x^2 - 4$

C) $-4 + \frac{1}{4}x^2$

D) $x^2 + 8x + 12$

109) Evaluate the below function.

$$p(x) = 4x + 1; \text{ Find } p(-2 + x)$$

A) $17 - 4x$ B) $-11 - 4x$

C) $-7 + 4x$ D) $16x + 1$

110) Evaluate the below function.

$$h(x) = x^2 + 1; \text{ Find } h(x + 2)$$

A) $x^4 + 1$ B) $4x^2 + 1$

C) $16x^2 + 1$ D) $x^2 + 4x + 5$

111) Evaluate the below function.

$$k(x) = x^2 + 5x; \text{ Find } k(-3x)$$

A) $4x^2 - 10x$ B) $-6 - x + x^2$

C) $9x^2 - 15x$ D) $x^2 - 5x$

112) Evaluate the below function.

$$f(n) = 3n - 2; \text{ Find } f(n + 4)$$

A) $3n - 14$ B) $-3n - 2$

C) $3n - 5$ D) $3n + 10$

113) Evaluate the below function.

$$g(n) = n^3 - 2n^2; \text{ Find } g(t - 4)$$
A) $t^3 - 14t^2 + 64t - 96$
B) $-t^3 - 2t^2$
C) $\dfrac{1}{8}t^3 - \dfrac{1}{2}t^2$
D) $-27t^3 - 18t^2$

114) Evaluate the below function.

$$h(n) = 2n + 1; \text{ Find } h(n + 2)$$

A) $2n + 5$ B) $2n - 3$

C) $2n - 7$ D) $2n + 7$

115) Evaluate the below function.

$$h(n) = 2n^2 + 3n; \text{ Find } h(2n)$$
A) $2n^4 + 3n^2$
B) $2n^2 + 15n + 27$
C) $2n^2 - 13n + 20$
D) $8n^2 + 6n$

116) Evaluate the below function.

$$k(x) = 3x - 4; \text{ Find } k(3 + x)$$

A) $5 + 3x$ B) $12x - 4$

C) $-3x - 4$ D) $3x - 13$

117) Evaluate the below function.

$$f(n) = n^3 - 2n; \text{ Find } f(n + 1)$$

A) $n^3 + 3n^2 + n - 1$
B) $n^3 - 6n^2 + 10n - 4$
C) $-n^3 + 2n$
D) $-4 - 10n - 6n^2 - n^3$

118) Evaluate the below function.

$$g(n) = 5^{-n} - 1; \text{ Find } g(n - 2)$$

A) $5^{-n+2} - 1$ B) $5^{-n-4} - 1$

C) $5^n - 1$ D) $5^{-\frac{1}{3}n} - 1$

119) Evaluate the below function.

$$g(x) = 3^x; \text{ Find } g(x - 2)$$

A) 3^{x-2} B) 3^{2x}

C) $3^{\frac{1}{4}x}$ D) 3^{x-4}

120) Evaluate the below function.

$$p(x) = 5^{2x} + 3; \text{ Find } p(x - 3)$$

A) $5^{\frac{2}{3}x} + 3$ B) $5^{2x-6} + 3$

C) $5^{6x} + 3$ D) $\dfrac{3 \cdot 5^{6x} + 1}{5^{6x}}$

        www.math-knots.com | www.a4ace.com

121) Evaluate the below function.

$$p(x) = -2|x + 1|; \text{ Find } p(4 - x)$$

A) $-2|3x + 1|$ B) $-2|5 - x|$

C) $-2|x - 1|$ D) $-2|x - 2|$

122) Evaluate the below function.

$$p(n) = -3 \cdot 3^n + 2; \text{ Find } p(x + 1)$$

A) $-3^{x^2 + 1} + 2$ B) $\dfrac{2 \cdot 3^{3x} - 3}{3^{3x}}$

C) $-3^{4 + x} + 2$ D) $-3^{x + 2} + 2$

123) Evaluate the below function.

$$f(x) = 5^{-x}; \text{ Find } f(-2x)$$

A) $\dfrac{1}{5^{2x}}$ B) $\dfrac{1}{5^{3x}}$

C) 5^{2x} D) $5^{-1 - x}$

124) Evaluate the below function.

$$f(x) = 2 \cdot 3^{-x}; \text{ Find } f(x^2)$$

A) $2 \cdot 3^{-1 + x}$ B) $\dfrac{2}{3^{x^2}}$

C) $2 \cdot 3^{-x - 1}$ D) $\dfrac{2}{3^{4x}}$

125) Evaluate the below function.

$$w(n) = -3|n|; \text{ Find } w(-4n)$$

A) $-3|-3n|$ B) $-3|-4n|$

C) $-3|1 - n|$ D) $-3|n + 1|$

126) Evaluate the below function.

$$f(n) = 2|n|; \text{ Find } f\left(\dfrac{n}{3}\right)$$

A) $2n^2$ B) $2\left|\dfrac{1}{3}n\right|$

C) $2\left|\dfrac{1}{4}n\right|$ D) $2|n - 1|$

127) Evaluate the below function.

$$g(n) = |-2n| + 1; \text{ Find } g(-4n)$$

A) $|-6 - 2n| + 1$

B) $|-2n + 2| + 1$

C) $|8n| + 1$

D) $|-8 - 2n| + 1$

128) Evaluate the below function.

$$g(n) = 3^n; \text{ Find } g(3n)$$

A) $\dfrac{1}{3^{3n}}$ B) $3^{1 - n}$

C) 3^{3n} D) $3^{n + 1}$

129) Evaluate the below function.

$$h(t) = |3t - 1|; \text{ Find } h(n + 2)$$

A) $|3n + 5|$ B) $|9n - 1|$

C) $|3n - 1|$ D) $|-9n - 1|$

130) Evaluate the below function.

$$g(n) = 3|n| - 1; \text{ Find } g(n - 3)$$

A) $3|n - 2| - 1$

B) $3|4 + n| - 1$

C) $3|n - 3| - 1$

D) $3|3n| - 1$

131) Evaluate the below function.

$$w(a) = -3|a| \; ; \text{ Find } w(a^2)$$

A) $-3|4+a|$ B) $-3a^2$

C) $-3|-3a|$ D) $-3|-2a|$

132) Evaluate the below function.

$$g(x) = |x| - 2; \text{ Find } g\left(\frac{x}{2}\right)$$

A) $|3x| - 2$ B) $x^2 - 2$

C) $\left|\frac{1}{2}x\right| - 2$ D) $|3+x| - 2$

133) Evaluate the below function.

$$f(n) = |n| + 3; \text{ Find } f(-2y)$$

A) $|3y| + 3$ B) $|-2y| + 3$

C) $|y| + 3$ D) $|y-4| + 3$

134) Evaluate the below function.

$$g(t) = 5^{2t-1}; \text{ Find } g(-3-t)$$

A) 5^{6t-1} B) 5^{-6t-1}

C) 5^{2t+7} D) 5^{-7-2t}

135) Evaluate the below function.

$$g(n) = 2^n; \text{ Find } g(n^2)$$

A) 2^{3n} B) 2^{-1+n}

C) $\dfrac{1}{2^{2n}}$ D) 2^{n^2}

136) Evaluate the below function.

$$p(t) = 3^{3t}; \text{ Find } p(2t)$$

A) $\dfrac{1}{3^{6t}}$ B) 3^{6t}

C) 3^{6+3t} D) 3^{9t}

137) The diagram below represents a relation. Identify the set of order pairs, domain and range and state if the relation is function or not

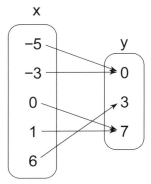

A) $\{(-5, 0), (-3, 0), (1, 7), (5, 5), (6, 3)\}$
Domain: $\{-5, -3, 1, 5, 6\}$
Range: $\{0, 3, 5, 7\}$
The relation is a function.

B) $\{(-3, 5), (1, 7), (3, 5), (5, -3), (6, 3)\}$
Domain: $\{-3, 1, 3, 5, 6\}$
Range: $\{-3, 3, 5, 7\}$
The relation is a function.

C) $\{(-5, 0), (-4, 5), (-4, -1), (0, 7), (6, 3)\}$
Domain: $\{-5, -4, 0, 6\}$
Range: $\{-1, 0, 3, 5, 7\}$
The relation is not a function.

D) $\{(-5, 0), (-3, 0), (0, 7), (1, 7), (6, 3)\}$
Domain: $\{-5, -3, 0, 1, 6\}$
Range: $\{0, 3, 7\}$
The relation is a function.

138) The diagram below represents a relation. Identify the set of order pairs, domain and range and state if the relation is function or not

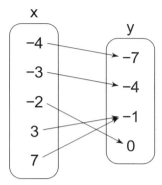

A) { (−4, −7), (−3, −4), (−2, 0), (3, −1), (7, −1) } Domain: $\{-4, -3, -2, 3, 7\}$

Range: $\{-7, -4, -1, 0\}$

The relation is not a function.

B) { (−4, −7), (−2, 0), (0, 7), (3, −1), (7, −1) }

Domain: $\{-4, -2, 0, 3, 7\}$

Range: $\{-7, -1, 0, 7\}$

The relation is a function.

C) { (−5, −2), (−4, −7), (−3, −4), (7, −2), (7, −1) } Domain: $\{-5, -4, -3, 7\}$

Range: $\{-7, -4, -2, -1\}$

The relation is not a function.

D) { (−6, 3), (−4, −7), (−3, −4), (−2, 0), (4, 4) } Domain: $\{-6, -4, -3, -2, 4\}$

Range: $\{-7, -4, 0, 3, 4\}$

The relation is a function.

139) The diagram below represents a relation. Identify the set of order pairs, domain and range and state if the relation is function or not

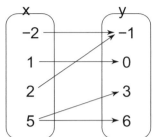

A) { (−2, −1), (1, 0), (2, −1), (5, 6), (5, 3) }

Domain: $\{-2, 1, 2, 5\}$

Range: $\{-1, 0, 3, 6\}$

The relation is not a function.

B) { (−4, 2), (−2, −1), (1, 0), (5, 6), (5, 3) }

Domain: $\{-4, -2, 1, 5\}$

Range: $\{-1, 0, 2, 3, 6\}$

The relation is not a function.

C) { (−2, 1), (−2, −1), (2, −1), (5, 6), (5, 3) }

Domain: $\{-2, 2, 5\}$

Range: $\{-1, 1, 3, 6\}$

The relation is not a function.

D) { (−6, −1), (−2, −1), (1, 0), (2, −1), (5, 6) }

Domain: $\{-6, -2, 1, 2, 5\}$

Range: $\{-1, 0, 6\}$

The relation is a function.

www.math-knots.com | www.a4ace.com

Algebra 1

140) The diagram below represents a relation. Identify the set of order pairs, domain and range and state if the relation is function or not

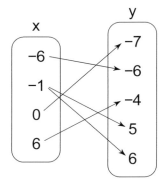

A) { (−6, −6), (−1, 6), (−1, 5), (0, −7), (6, −4) }

   Domain: {−6, −1, 0, 6}

   Range: {−7, −6, −4, 5, 6}

   The relation is not a function.

B) { (−6, −6), (−4, 1), (−1, 6), (0, −7), (6, −4) }

   Domain: {−6, −4, −1, 0, 6}

   Range: {−7, −6, −4, 1, 6}

   The relation is a function.

C) { (−6, 6), (−6, −6), (−1, 6), (−1, 5), (6, 2) }

   Domain: {−6, −1, 6}

   Range: {−6, 2, 5, 6}

   The relation is not a function.

D) { (−6, 4), (−5, −5), (4, 7), (6, −4), (7, 0) }

   Domain: {−6, −5, 4, 6, 7}

   Range: {−5, −4, 0, 4, 7}

   The relation is a function.

141) The diagram below represents a relation. Identify the set of order pairs, domain and range and state if the relation is function or not

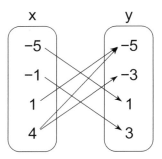

A) { (−7, 1), (−5, 1), (−2, 6), (1, −5), (4, −5) }

   Domain: {−7, −5, −2, 1, 4}

   Range: {−5, 1, 6}

   The relation is a function.

B) { (−5, 1), (−1, 3), (1, −5), (4, −5), (7, −3) }

   Domain: {−5, −1, 1, 4, 7}

   Range: {−5, −3, 1, 3}

   The relation is a function.

C) { (−5, 1), (−1, 3), (1, −5), (4, −3), (4, −5) }

   Domain: {−5, −1, 1, 4}

   Range: {−5, −3, 1, 3}

   The relation is not a function.

D) { (−6, 4), (−5, 1), (−1, 3), (1, −5), (4, −5) }

   Domain: {−6, −5, −1, 1, 4}

   Range: {−5, 1, 3, 4}

   The relation is a function.

www.math-knots.com | www.a4ace.com

142) The graph below represents a relation. Identify the set of order pairs, domain, range, mapping diagram and state if the relation is function or not

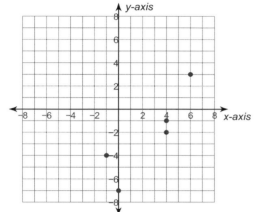

A)

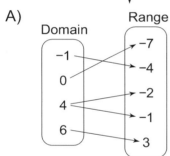

$\{(-1, -4), (0, -7), (4, -2), (4, -1), (6, 3)\}$
Domain: $\{-1, 0, 4, 6\}$
Range: $\{-7, -4, -2, -1, 3\}$
The relation is not a function.

B)

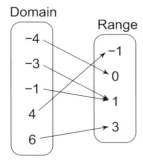

$\{(-4, 0), (-3, 1), (-1, 1), (4, -1), (6, 3)\}$
Domain: $\{-4, -3, -1, 4, 6\}$
Range: $\{-1, 0, 1, 3\}$
The relation is a function.

143) The graph below represents a relation. Identify the set of order pairs, domain, range, mapping diagram and state if the relation is function or not

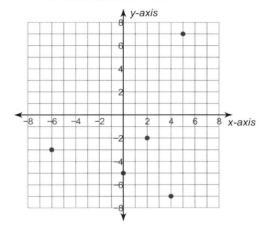

A)

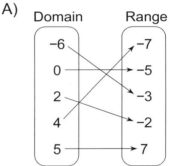

$\{(-6, -3), (0, -5), (2, -2), (4, -7), (5, 7)\}$
Domain: $\{-6, 0, 2, 4, 5\}$
Range: $\{-7, -5, -3, -2, 7\}$
The relation is a function.

B)

$\{(-6, -3), (-1, 3), (2, -2), (4, -7), (5, 7)\}$
Domain: $\{-6, -1, 2, 4, 5\}$
Range: $\{-7, -3, -2, 3, 7\}$
The relation is a function.

Algebra 1

144) The graph below represents a relation. Identify the set of order pairs, domain, range, mapping diagram and state if the relation is function or not

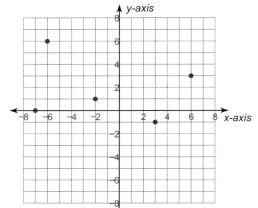

A)   Domain

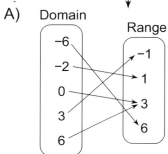

{ (−6, 6), (−2, 1), (0, 3), (3, −1), (6, 3) }

Domain: {−6, −2, 0, 3, 6}

Range: {−1, 1, 3, 6}

The relation is a function.

B)   Domain   Range

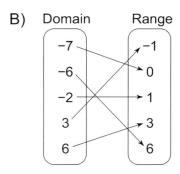

{ (−7, 0), (−6, 6), (−2, 1), (3, −1), (6, 3) }

Domain: {−7, −6, −2, 3, 6}

Range: {−1, 0, 1, 3, 6}

The relation is a function.

145) The graph below represents a relation. Identify the set of order pairs, domain, range, mapping diagram and state if the relation is function or not

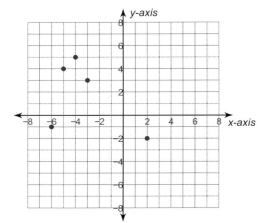

A)
   Domain   Range

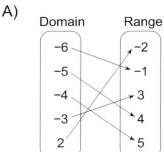

{ (−6, −1), (−5, 4), (−4, 5), (−3, 3), (2, −2) }
Domain: {−6, −5, −4, −3, 2}
Range: {−2, −1, 3, 4, 5}
The relation is a function.

B)
   Domain   Range

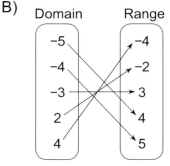

{ (−5, 4), (−4, 5), (−3, 3), (2, −2), (4, −4) }
Domain: {−5, −4, −3, 2, 4}
Range: {−4, −2, 3, 4, 5}
The relation is a function.

146) The graph below represents a relation. Identify the set of order pairs, domain, range, mapping diagram and state if the relation is function or not

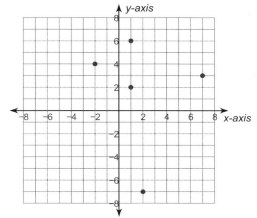

A)

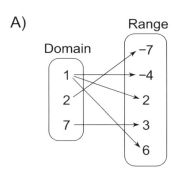

{(1, 6), (1, 2), (1, −4), (2, −7), (7, 3)}
Domain: {1, 2, 7}
Range: {−7, −4, 2, 3, 6}
The relation is not a function.

B)

Range
Domain
−2
1
2
7
−7
2
3
4
6

{(−2, 4), (1, 2), (1, 6), (2, −7), (7, 3)}
Domain: {−2, 1, 2, 7}
Range: {−7, 2, 3, 4, 6}
The relation is not a function.

147) The set below represents a relation. Identify the domain, range, mapping diagram and state if the relation is function or not

{(−6, −3), (−4, 6), (0, 1), (3, 2), (4, 1), (4, −4)}

A)

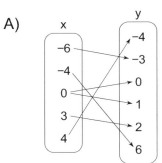

Domain: {−6, −4, 0, 3, 4}
Range: {−4, −3, 0, 1, 2, 6}
The relation is not a function.

B)

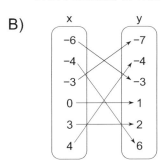

Domain: {−6, −4, −3, 0, 3, 4}
Range: {−7, −4, −3, 1, 2, 6}
The relation is a function.

C)

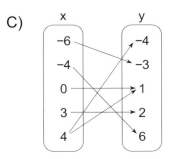

Domain: {−6, −4, 0, 3, 4}
Range: {−4, −3, 1, 2, 6}
The relation is not a function.

www.math-knots.com | www.a4ace.com

Algebra 1

148) The set below represents a relation. Identify the domain, range, mapping diagram and state if the relation is function or not

$\{(-5, -3), (-3, -3), (-1, -3), (0, 7), (0, -4), (2, 0)\}$

A)

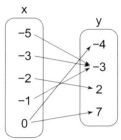

Domain: $\{-5, -3, -2, -1, 0\}$
Range: $\{-4, -3, 2, 7\}$
The relation is not a function.

B)

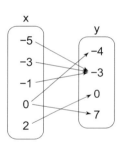

Domain: $\{-5, -3, -1, 0, 2\}$
Range: $\{-4, -3, 0, 7\}$
The relation is not a function.

C)

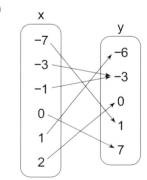

Domain: $\{-7, -3, -1, 0, 1, 2\}$
Range: $\{-6, -3, 0, 1, 7\}$
The relation is a function.

149) The set below represents a relation. Identify the domain, range, mapping diagram and state if the relation is function or not

$\{(-3, -3), (-3, -6), (-3, 0),$
$(-2, 2), (7, -4), (7, 0)\}$

A)

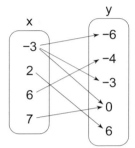

Domain: $\{-3, 2, 6, 7\}$
Range: $\{-6, -4, -3, 0, 6\}$
The relation is not a function.

B)

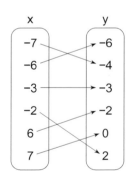

Domain: $\{-7, -6, -3, -2, 6, 7\}$
Range: $\{-6, -4, -3, -2, 0, 2\}$
The relation is a function.

C)

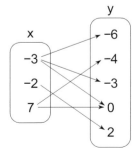

Domain: $\{-3, -2, 7\}$
Range: $\{-6, -4, -3, 0, 2\}$
The relation is not a function.

150) The set below represents a relation. Identify the domain, range, mapping diagram and state if the relation is function or not

$$\{(-7, 3), (-6, -5), (-4, 2), (-3, 5), (5, -3), (5, -7)\}$$

A)

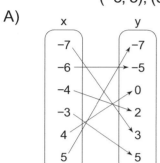

Domain: $\{-7, -6, -4, -3, 4, 5\}$
Range: $\{-7, -5, 0, 2, 3, 5\}$
The relation is a function.

B)

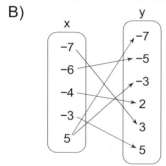

Domain: $\{-7, -6, -4, -3, 5\}$
Range: $\{-7, -5, -3, 2, 3, 5\}$
The relation is not a function.

C)

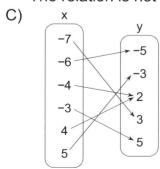

Domain: $\{-7, -6, -4, -3, 4, 5\}$
Range: $\{-5, -3, 2, 3, 5\}$
The relation is a function.

151) The table below represents a relation. Identify the domain, range and state if the relation is function or not

x	y
-3	0
1	-1
2	-7
2	5
3	6

A) Domain: $\{-3, 1, 2, 3\}$
Range: $\{-7, -1, 0, 5, 6\}$
The relation is not a function.

B) Domain: $\{-3, 0, 1, 2, 3\}$
Range: $\{-1, 0, 5, 6\}$
The relation is a function.

C) Domain: $\{-6, -3, -2, 1, 6\}$
Range: $\{-4, -1, 0, 5\}$
The relation is a function.

152) The table below represents a relation. Identify the domain, range and state if the relation is function or not

x	y
-3	6
-3	-3
0	-1
0	-3
3	2

A) Domain: $\{-3, 0, 3\}$
Range: $\{-3, -1, 2, 6\}$
The relation is not a function.

B) Domain: $\{-7, -3, 0, 3\}$
Range: $\{-3, -1, 2\}$
The relation is not a function.

C) Domain: $\{-3, 0, 2, 3, 7\}$
Range: $\{-3, 1, 2, 5, 6\}$
The relation is a function.

www.math-knots.com | www.a4ace.com

153) The table below represents a relation. Identify the domain, range and state if the relation is function or not

x	y
−6	−7
−3	−7
−1	5
2	−1
3	−4

A) Domain: $\{-6, -3, -1, 2, 3\}$
Range: $\{-7, -4, -1, 5\}$
The relation is a function.

B) Domain: $\{-6, -3, -1, 2, 4\}$
Range: $\{-7, -1, 2, 5\}$
The relation is a function.

C) Domain: $\{-3, -2, -1, 2, 3\}$
Range: $\{-7, -4, -1, 3, 5\}$
The relation is a function.

154) The table below represents a relation. Identify the domain, range and state if the relation is function or not

x	y
−2	−4
0	6
0	7
3	−2
4	6

A) Domain: $\{-2, 0, 3\}$
Range: $\{-4, -2, 3, 6, 7\}$
The relation is not a function.

B) Domain: $\{-2, 0, 3, 4\}$
Range: $\{-4, -2, 6, 7\}$
The relation is not a function.

C) Domain: $\{-3, -2, 0, 3, 4\}$
Range: $\{-4, -2, 6, 7\}$
The relation is a function.

155) The table below represents a relation. Identify the domain, range and state if the relation is function or not

x	y
−7	−7
−6	3
1	−6
3	−7
7	−7

A) Domain: $\{-7, -3, 1, 3\}$
Range: $\{-7, -6, -1\}$
The relation is not a function.

B) Domain: $\{-7, -6, 1, 7\}$
Range: $\{-7, -6, -3, 3\}$
The relation is not a function.

C) Domain: $\{-7, -6, 1, 3, 7\}$
Range: $\{-7, -6, 3\}$
The relation is a function.

156) The table below represents a relation. Identify the domain, range and state if the relation is function or not

x	y
−7	−1
0	6
1	−3
3	5
7	−3

A) Domain: $\{-7, 1, 3, 6\}$
Range: $\{-4, -3, -2, -1, 5\}$
The relation is not a function.

B) Domain: $\{-7, 1, 3, 7\}$
Range: $\{-4, -3, -1, 1\}$
The relation is not a function.

C) Domain: $\{-7, 0, 1, 3, 7\}$
Range: $\{-3, -1, 5, 6\}$
The relation is a function.

1) Evaluate the below function.

$$p(a) = a^2 - a$$

Given : { – 5 , 0 , 3, 6, 7 }

Domain :

Range :

2) Evaluate the below function.

$$h(n) = 4n - 1$$

Given : { – 8 , – 6 , – 1, 0, 10, 11 }

Domain :

Range :

3) Evaluate the below function.

$$w(x) = x^2 - 5 + x$$

Given : { – 4 , – 3 , – 2, 1, 5 }

Domain :

Range :

4) Evaluate the below function.

$$f(t) = t^3 - 5 - 2t$$

Given : { – 4 , – 2 , – 1, 1, 2, 5 }

Domain :

Range :

5) Evaluate the below function.

$$p(n) = 4n + 4$$

Given : { – 9 , – 7 , – 2, 1, 6, 10 }

Domain :

Range :

6) Evaluate the below function.

$$f(x) = 4x + 4$$

Given : { – 12 , – 15 , – 5, 10, 20, 11 }

Domain :

Range :

7) Evaluate the below function.

$$f(x) = x^2 + 2$$

Given : { – 8 , – 4 , – 1, 2, 4, 10 }

Domain :

Range :

8) Evaluate the below function.

$$p(t) = t^2 + 1$$

Given : { – 11 , – 10 , – 5, 10, 12, 13 }

Domain :

Range :

9) Evaluate the below function.

$$w(a) = a - 2$$

Given : { – 18 , – 15 , – 11, 12, 17, 80 }

Domain :

Range :

10) Evaluate the below function.

$$g(t) = -t + 4$$

Given : { – 25 , – 33 , – 45, 50, 64, 77 }

Domain :

Range :

11) Evaluate the below function.

$$g(x) = x^3 - 3x$$

Given : { − 4 , − 3 , − 2, 1, 2 , 5 }

Domain :

Range :

12) Evaluate the below function.

$$h(t) = t^3 + 1 - 2t$$

Given : { − 4 , − 2 , − 1, 1, 2, 5 }

Domain :

Range :

13) Evaluate the below function.

$$p(t) = 3t + 3$$

Given : { − 11 , − 14 , − 2, 1, 7 , 10 }

Domain :

Range :

14) Evaluate the below function.

$$g(n) = 3n - 5$$

Given : { − 12 , − 15 , − 5, 10, 20, 11 }

Domain :

Range :

15) Evaluate the below function.

$$f(x) = x^3 + 5$$

Given : { − 8 , − 4 , − 1, 2, 4 }

Domain :

Range :

16) Evaluate the below function.

$$f(x) = x^3 + 3x \; ; \; \text{Find } f(-3-x)$$

Given : { − 4 , − 10 , − 5, 1, 2, 3 }

Domain :

Range :

17) Evaluate the below function.

$$h(n) = 2n + 2$$

Given : { − 9 , − 7 , − 2, 1, 6, 10 }

Domain :

Range :

18) Evaluate the below function.

$$f(t) = 2t + 3$$

Given : { − 12 , − 15 , − 5, 10, 20, 11 }

Domain :

Range :

19) Evaluate the below function.

$$w(a) = a^3 - 2a^2$$

Given : { − 5 , − 3 , − 2, 1, 2, 5 }

Domain :

Range :

20) Evaluate the below function.

$$w(x) = x^2 + 2$$

Given : { − 5 , − 2 , − 1, 1, 2, 5 }

Domain :

Range :

21) Evaluate the below function.

$$h(t) = t^3 + 4t^2$$

Given : { − 4 , − 3 , − 2 , 1 , 2 , 5 }

Domain :

Range :

22) Evaluate the below function.

$$h(a) = 3a + 4$$

Given : { − 4 , − 2 , − 1 , 1, 2, 5 }

Domain :

Range :

23) Evaluate the below function.

$$g(a) = 2a^2 - 4 - 2a$$

Given : { − 4 , − 2 , 0, 1, 2 , 3 }

Domain :

Range :

24) Evaluate the below function.

$$g(a) = a + 4$$

Given : { − 4 , − 2 , − 1, 1, 2, 5 }

Domain :

Range :

25) Evaluate the below function.

$$f(a) = a^2 - 5$$

Given : { − 4 , − 3 , − 2, 1, 2, 5 }

Domain :

Range :

26) Evaluate the below function.

$$g(x) = 2x + 5$$

Given : { − 4 , − 2 , − 1, 1, 2 , 5 }

Domain :

Range :

27) Evaluate the below function.

$$h(n) = n^2 + 1$$

Given : { − 5 , − 1 , − 2, 1, 0, 5 }

Domain :

Range :

28) Evaluate the below function.

$$g(x) = x^2 - 5$$

Given : { − 3 , − 2 , − 1, 1, 6, 7 }

Domain :

Range :

29) Evaluate the below function.

$w(x) = 2x$

Given : $\{-11, -14, -2, 1, 7, 10\}$

Domain :

Range :

30) Evaluate the below function.

$f(x) = x^2 + 4$

Given : $\{-12, -15, -5, 10, 20, 11\}$

Domain :

Range :

31) Evaluate the below function.

$g(n) = n^3 - 5$

$f(n) = 4n - 2$

Find $g(-5) + f(-5)$

A) -152 B) 138

C) -12 D) 1

32) Evaluate the below function.

$f(x) = n^3 - 5$

$g(x) = 4x - 5$

Find $(f + g)(-3)$

A) 3   B) 12

C) 9   D) -49

33) Evaluate the below function.

$f(x) = x^2 - 2$

$g(x) = 3x + 1$

Find $(f + g)(-9)$

A) 17   B) 107

C) 53   D) 87

34) Evaluate the below function.

$g(a) = 3a^2 + 2a$

$f(a) = a - 5$

Find $(g + f)(-6)$

A) 163   B) 1

C) 121   D) 85

35) Evaluate the below function.

$f(n) = n - 3$

$g(n) = n^3 - 3n$

Find $f(-1) + g(-1)$

A) -2 B) -24

C) -4 D) 112

36) Evaluate the below function.

$h(a) = -a^3 - a$

$g(a) = 4a - 2$

Find $(h + g)(-3)$

A) -2 B) -36

C) -20 D) -14

37) Evaluate the below function.

$$g(x) = 4x + 3$$
$$h(x) = 3x + 2$$
Find $(g + h)(-9)$

A) -58 B) 26

C) 61   D) 68

38) Evaluate the below function.

$$h(x) = x^2 - 1$$
$$g(x) = -x + 5$$
Find $h(-9) + g(-9)$

A) 4   B) 76

C) 10   D) 94

39) Evaluate the below function.

$$g(x) = 4x - 3$$
$$f(x) = -2x^2 - 5$$
Find $g(-6) + f(-6)$

A) -168 B) -104

C) -56 D) -24

40) Evaluate the below function.

$$g(a) = a^3 - 3a$$
$$h(a) = 2a + 5$$
Find $(g + h)(4)$

A) -55 B) 65

C) 40   D) 5

41) Evaluate the below function.

$$g(t) = 3t - 4$$
$$f(t) = -t^2 + 5 + t$$
Find $(g + f)(2)$

A) 1   B) 5

C) -95 D) -11

42) Evaluate the below function.

$$g(a) = 2a + 4$$
$$h(a) = a^3 - 2a^2$$
Find $(g + h)(5)$

A) 89   B) -181

C) 44   D) -1

43) Evaluate the below function.

$$g(a) = 2a - 1$$
$$f(a) = a + 5$$
Find $(g + f)(-8)$

A) 7   B) -14

C) -20 D) 28

44) Evaluate the below function.

$$f(x) = x^2 + 4$$
$$g(x) = 3x - 4$$
Find $f(6) + g(6)$

A) 40   B) 18

C) 54   D) 70

www.math-knots.com | www.a4ace.com

45) Evaluate the below function.

$$g(t) = t - 3$$
$$h(t) = t^2 + 1$$

Find $(g + h)(-3)$

A) 28   B) 10

C) 4   D) 0

46) Evaluate the below function.

$$h(a) = 2a - 4$$
$$g(a) = 2a^2 - 3$$

Find $h(-1) + g(-1)$

A) – 7   B) – 3

C) – 25   D) 2

47) Evaluate the below function.

$$g(n) = 3n + 2$$
$$f(n) = n^2 + n$$

Find $(g + f)(-3)$

A) – 2   B) – 1

C) 23   D) 119

48) Evaluate the below function.

$$g(a) = 3t - 4$$
$$h(a) = t^2 + 3t$$

Find $(g + h)(-3)$

A) 51   B) 36

C) – 13   D) 23

49) Evaluate the below function.

$$h(x) = -2x - 2$$
$$g(x) = x^2 + 2x$$

Find $(h + g)(-3)$

A) 131   B) 98

C) 7   D) – 2

50) Evaluate the below function.

$$h(n) = n + 5$$
$$g(n) = 3n^2 + 1$$

Find $h(-1) + g(-1)$

A) 108   B) 8

C) 6   D) 10

51) Evaluate the below function.

$$g(x) = x^3 + 4$$
$$h(x) = x + 5$$

Find $(g + h)(-2)$

A) – 1   B) 8

C) 3   D) 19

52) Evaluate the below function.

$$h(x) = -x + 1$$
$$g(x) = 3x - 5$$

Find $h(2) + g(2)$

A) – 8   B) 10

C) 0   D) 39

53) Evaluate the below function.

$$h(t) = 3t - 3$$
$$g(t) = t^3 - 4t$$

Find $(h + g)(0)$

A) -5 B) -1

C) -2 D) -3

54) Evaluate the below function.

$$g(x) = x^3 - 5x$$
$$h(x) = x - 2$$

Find $g(1) + h(1)$

A) 1   B) 46

C) -9 D) -5

55) Evaluate the below function.

$$g(x) = x^3 - x$$
$$f(x) = 2x + 5$$

Find $g(4) + f(4)$

A) 3   B) -6

C) 73   D) -63

56) Evaluate the below function.

$$f(x) = -4x$$
$$g(x) = x^2 + 1$$

Find $f(10) - g(10)$

A) -1 B) 141

C) -141 D) 61

57) Evaluate the below function.

$$f(x) = x^2 + 5$$
$$g(x) = x + 2$$

Find $(f - g)(-2)$

A) -5 B) -9

C) 9   D) 113

58) Evaluate the below function.

$$h(x) = x^2 - 1$$
$$g(x) = -2x - 3$$

Find $h(-9) - g(-9)$

A) -101 B) -65

C) 65   D) 2

59) Evaluate the below function.

$$f(n) = -n^3 - 4$$
$$g(n) = 3n + 4$$

Find $f(3) - g(3)$

A) -44 B) 28

C) -8 D) 44

60) Evaluate the below function.

$$f(a) = 4a$$
$$g(a) = 3a + 3$$

Find $(f - g)(-4)$

A) -7 B) 7

C) -1 D) -5

61) Evaluate the below function.

$$g(a) = -3a + 5$$
$$h(a) = -3a - 2$$

Find $g(2) - h(2)$

A) 7   B) -7

C) -19 D) -3

62) Evaluate the below function.

$$g(x) = x^3 + 4x^2$$
$$h(x) = 4x$$

Find $(g - h)(-3)$

A) 21   B) 16

C) -21 D) -51

63) Evaluate the below function.

$$h(t) = t + 3$$
$$g(t) = t^2 + 5t$$

Find $h(9) - g(9)$

A) 42   B) 114

C) -114 D) -2

64) Evaluate the below function.

$$f(x) = 2x - 5$$
$$g(x) = 4x + 5$$

Find $(f - g)(9)$

A) -10 B) 28

C) -8 D) -28

65) Evaluate the below function.

$$g(n) = -n + 4$$
$$h(n) = 3n - 4$$

Find $(g - h)(3)$

A) 4   B) 28

C) -20 D) -4

66) Evaluate the below function.

$$f(x) = 4x$$
$$g(x) = x^2 - 3 - x$$

Find $(f - g)(1)$

A) 3   B) -7

C) -21 D) 7

67) Evaluate the below function.

$$g(x) = -x^2 + 5x$$
$$h(x) = 2x - 3$$

Find $g(2) - h(2)$

A) -25 B) -5

C) 5   D) -7

68) Evaluate the below function.

$$f(n) = 2n$$
$$g(n) = 4n + 4$$

Find $(f - g)(5)$

A) -6 B) -14

C) 10   D) 14

www.math-knots.com | www.a4ace.com

69) Evaluate the below function.

$$h(t) = t^2 - 2$$
$$g(t) = 4t + 4$$

Find $h(-1) - g(-1)$

A) 90   B) -1

C) 9   D) 1

70) Evaluate the below function.

$$g(x) = 3x^2 + 5$$
$$h(x) = -2x - 2$$

Find $(g - h)(5)$

A) 92   B) -92

C) 7   D) -72

71) Evaluate the below function.

$$f(t) = t - 3$$
$$g(t) = 3t - 4$$

Find $(f - g)(-7)$

A) -15 B) -13

C) 38   D) 15

72) Evaluate the below function.

$$g(n) = n^3 + n^2$$
$$h(n) = 3n - 5$$

Find $g(1) - h(1)$

A) 4   B) -8

C) 8   D) -4

73) Evaluate the below function.

$$f(n) = n - 5$$
$$g(n) = n^2 - 4$$

Find $(f - g)(-7)$

A) 43   B) -57

C) -1 D) 57

74) Evaluate the below function.

$$f(t) = -2t - 4$$
$$g(t) = -t - 1$$

Find $f(-5) - g(-5)$

A) 8   B) -2

C) -12 D) 2

75) Evaluate the below function.

$$g(x) = 3x + 3$$
$$h(x) = 4x - 4$$

Find $(g - h)(9)$

A) -2 B) 2

C) -16 D) 17

76) Evaluate the below function.

$$g(n) = n^2 - 1$$
$$h(n) = 4n - 2$$

Find $(g - h)(3)$

A) -22 B) 2

C) -2 D) 33

77) Evaluate the below function.

$$g(x) = x^2 + 4$$
$$h(x) = x + 4$$
Find $(g + h)(x)$

A) $x^2 + x + 8$ B) $-x^3 - 3x + 2$

C) $x^2 - x + 8$ D) $x^2 - x - 6$

78) Evaluate the below function.

$$g(x) = x^2 - 4$$
$$h(x) = 4x$$
Find $(g + h)(x)$

A) $x^3 + x - 1$ B) $x^2 + x + 4$

C) $x^2 + 8x - 1$ D) $x^2 + 4x - 4$

79) Evaluate the below function.

$$f(x) = -4x - 3$$
$$g(x) = 3x + 4$$
Find $f(x) + g(x)$

A) $-x + 1$ B) $2x^2 - x - 4$

C) $x + 1$ D) $-6x + 7$

80) Evaluate the below function.

$$g(n) = n - 3$$
$$f(n) = 2n + 4$$
Find $(g + f)(n)$

A) $-5n + 1$ B) $-3n + 1$

C) $n^2 + n - 2$ D) $3n + 1$

81) Evaluate the below function.

$$g(x) = -2x$$
$$h(x) = x^3 + 1$$
Find $(g + h)(x)$

A) $x^3 - 2x + 1$ B) $-x^3 + 4x$

C) $-x^3 + 2x + 1$ D) $-x^3 - 2x - 2$

82) Evaluate the below function.

$$f(n) = 3n - 3$$
$$g(n) = 2n - 2$$
Find $f(n) + g(n)$

A) $n^2 + 5$ B) $-5n - 5$

C) $-n^3 - 2n^2 - 2n - 3$ D) $5n - 5$

83) Evaluate the below function.

$$g(n) = n^3 + 5n^2$$
$$h(n) = 2n + 3$$
Find $(g + h)(n)$

A) $n^2 - 3n + 9$ B) $-n^3 + 5n^2 - 2n + 3$

C) n D) $n^3 + 5n^2 + 2n + 3$

84) Evaluate the below function.

$$h(a) = 4a - 4$$
$$g(a) = a^3 - 4a^2$$
Find $h(a) + g(a)$

A) $-a^3 - a + 6$ B) $-a^3 - 4a^2 - 4a - 4$

C) $-2a + 5$ D) $a^3 - 4a^2 + 4a - 4$

85) Evaluate the below function.

$$g(x) = x + 3$$
$$f(x) = 3x^3 + 3$$
Find $(g + f)(x)$

A) $-3x^3 - x + 6$ B) $3x^3 + x + 6$

C) $x^2 - x - 9$ D) $x^2 + 7$

86) Evaluate the below function.

$$f(a) = a^3 - 4a$$
$$g(a) = a - 2$$
Find $f(a) + g(a)$

A) $-2a^2 - 4a - 1$ B) $a^3 - 3a - 2$

C) $-2a^3 - 2a + 5$ D) $-a^3 + 3a - 2$

87) Evaluate the below function.

$$g(n) = 4n + 2$$
$$h(n) = 4n$$
Find $g(n) + h(n)$

A) $-5n - 7$ B) $-n^3 - 4n$

C) $-8n + 2$ D) $8n + 2$

88) Evaluate the below function.

$$g(a) = 2a^2 + 4$$
$$f(a) = -2a$$
Find $(g + f)(a)$

A) $2a^2 - 2a + 4$ B) $-2a^3 + a^2 - a - 3$

C) $a^2 - a + 3$ D) $2a^2 + 2a + 4$

89) Evaluate the below function.

$$h(x) = x^2 - 4x$$
$$g(x) = x + 1$$
Find $(h + g)(x)$

A) $3x - 3$ B) $x^3 - 3x + 5$

C) $x^2 - 3x + 1$ D) $-3x^2 + 3x + 5$

90) Evaluate the below function.

$$f(x) = x^3 + x^2$$
$$g(x) = x + 4$$
Find $(f + g)(x)$

A) $x^3 + x^2 + x + 4$
B) $-x^2 + x + 3$
C) $-x^3 - x^2 + x + 1$
D) $-x^3 + x^2 - x + 4$

91) Evaluate the below function.

$$g(n) = 4n$$
$$h(n) = n^2 + 5n$$
Find $(g + h)(n)$

A) $n^2 - 6n + 1$ B) $n^2 + 9n$

C) $n^2 - 9n$ D) $n^2 - 4n + 6$

92) Evaluate the below function.

$$f(x) = 4x + 1$$
$$g(x) = -x^2 + 3$$
Find $(f + g)(x)$

A) $-x^2 + 4x + 4$ B) $-x^2 - 4x + 4$

C) $x^2 + 2$ D) $x^2 - 2x + 2$

93) Evaluate the below function.

$$h(x) = x^2 + x$$
$$g(x) = 3x - 4$$
Find $h(x) + g(x)$

A) $x^2 - 4x - 4$ B) $x^2 + 4x - 4$

C) $3x + 1$ D) $x^2 - 3x + 10$

94) Evaluate the below function.

$$f(n) = -3n - 1$$
$$g(n) = 4n - 3$$
Find $(f + g)(n)$

A) $n - 4$ B) $-n - 4$

C) $-5n - 1$ D) $-3n - 7$

95) Evaluate the below function.

$$f(x) = 4x + 1$$
$$g(x) = -x^2 - x$$
Find $(f + g)(x)$

A) $-x^2 - 3x + 1$ B) $-x^2 + 3x + 1$

C) $-x^3 + 4x - 4$ D) $-3x^2 - x - 4$

96) Evaluate the below function.

$$g(t) = 3t - 4$$
$$h(t) = t^2 + 3$$
Find $g(t) + h(t)$

A) $-t^3 + 3t - 6$ B) $-t^3 - 3t - 1$

C) $t^3 + 3t - 1$ D) $t^2 - t - 5$

97) Evaluate the below function.

$$g(t) = 4t + 1$$
$$h(t) = t - 4$$
Find $g(t) + h(t)$

A) $t^2 - 6t - 4$ B) $5t - 3$

C) $-t^3 + t - 3$ D) $-5t - 3$

98) Evaluate the below function.

$$h(n) = n + 4$$
$$g(n) = -n + 5$$
Find $(h + g)(n)$

A) 9 B) $3n^3 - 9n + 1$

C) $n^2 - 4n - 3$ D) $-n + 7$

99) Evaluate the below function.

$$g(t) = 4t - 2$$
$$f(t) = t^3 - t^2$$
Find $g(t) + f(t)$

A) $-t^3 - t^2 + 4t - 2$ B) $t^3 - t^2 - 4t - 2$

C) $-5t + 7$ D) $-t^3 - t + 4$

100) Evaluate the below function.

$$f(n) = 4n + 1$$
$$g(n) = n^2 - 2$$
Find $f(n) + g(n)$

A) $n + 4n - 1$ B) $n^2 + 3n + 6$

C) $-n^2 - 3n - 4$ D) $n^2 - 4n - 1$

www.math-knots.com | www.a4ace.com

101) Evaluate the below function.

$$f(x) = x^3 - 2$$
$$g(x) = 4x - 2$$

Find $(f + g)(x)$

A) $x^3 + 4x - 4$ B) $x^2 + 3x - 1$

C) $2x^2 - 5x - 4$ D) $-x^3 - 4x - 4$

102) Evaluate the below function.

$$g(x) = 3x + 5$$
$$h(x) = 4x - 5$$

Find $(g - h)(x)$

A) $-x + 10$
B) $x - 10$
C) $-x^3 + 3x^2 + 4x - 1$
D) $-x - 10$

103) Evaluate the below function.

$$g(n) = 4n + 3$$
$$h(n) = 3n + 1$$

Find $g(n) - h(n)$

A) $-3n^2 + 3n - 1$ B) $-n - 2$

C) $-n + 2$ D) $n + 2$

104) Evaluate the below function.

$$h(t) = 2t - 2$$
$$g(t) = t - 3$$

Find $(h - g)(t)$

A) $t - 1$ B) $t^3 - 5t^2 - 2t - 1$

C) $-t - 1$ D) $t + 1$

105) Evaluate the below function.

$$f(n) = 3n + 1$$
$$g(n) = 4n + 1$$

Find $f(n) - g(n)$

A) $n^3 + 2$ B) $-n$

C) n D) $-3n^3 - 2n - 3$

106) Evaluate the below function.

$$h(x) = -2x^2 - 4$$
$$g(x) = 4x$$

Find $(h - g)(x)$

A) $-2x^2 + 4x - 4$
B) $2x^2 + 4x + 4$
C) $-2x^2 - 4x - 4$
D) $3x^3 - x + 1$

107) Evaluate the below function.

$$g(n) = 2n$$
$$f(n) = 3n - 4$$

Find $(g - f)(n)$

A) $-3n + 3$ B) $-n + 4$

C) $n - 4$ D) $-n - 4$

108) Evaluate the below function.

$$g(x) = 3x + 2$$
$$f(x) = 4x + 2$$

Find $g(x) - f(x)$

A) x B) $-x$

C) $x^2 + 1$ D) $2x^3 - 2x^2 - 4x - 5$

        www.math-knots.com | www.a4ace.com

109) Evaluate the below function.

$$g(a) = a^3 - 5a^2$$
$$f(a) = 3a + 1$$
Find $g(a) - f(a)$

A) $-a^3 + 5a^2 + 3a + 1$ B) $a^2 - 4a + 2$

C) $a^3 + 5a^2 - 3a + 1$ D) $a^3 - 5a^2 - 3a - 1$

110) Evaluate the below function.

$$f(x) = 4x + 1$$
$$g(x) = x^3 + x$$
Find $(f - g)(x)$

A) $x^3 - 3x - 1$

B) $x^2 - 2x$

C) $-x^3 + 3x - 1$

D) $-x^3 + 3x + 1$

111) Evaluate the below function.

$$g(t) = 3t - 4$$
$$f(t) = t^2 - 2 - t$$
Find $(g - f)(t)$

A) $t^3 - 3t + 4$ B) $-t^2 + 4t - 2$

C) $t^2 - 4t + 2$ D) $-t^2 - 4t - 2$

112) Evaluate the below function.

$$h(x) = x^2 - 5x$$
$$g(x) = 2x + 4$$
Find $h(x) - g(x)$

A) $-x^2 + 7x + 4$ B) $-x^2 + 2x$

C) $x^2 - 7x - 4$ D) $-x^2 - 7x + 4$

113) Evaluate the below function.

$$f(x) = 4x + 4$$
$$g(x) = 4x + 5$$
Find $f(x) - g(x)$

A) -1 B) $x^2 + 3x - 4$

C) $-2x^2 + 4x + 4$ D) 1

114) Evaluate the below function.

$$g(x) = 4x - 3$$
$$h(x) = x^3 + x^2$$
Find $g(x) - h(x)$

A) $x^3 + 6x + 3$

B) $-x^3 + x^2 + 4x + 3$

C) $-x^3 - x^2 + 4x - 3$

D) $x^3 + x^2 - 4x + 3$

115) Evaluate the below function.

$$h(a) = 3a^2 - a$$
$$g(a) = a + 1$$
Find $(h - g)(a)$

A) $-3a^2 + 2a + 1$ B) $-3a^2 - 2a + 1$

C) $3a^2 - 2a - 1$ D) $-2a^2 - 4a - 4$

116) Evaluate the below function.

$$h(x) = -4x - 2$$
$$g(x) = 4x + 1$$
Find $(h - g)(x)$

A) $-3x^2 - 6x - 4$ B) $-8x + 3$

C) $-8x - 3$ D) $8x + 3$

117) Evaluate the below function.

$$f(x) = x^3 + 1$$
$$g(x) = x - 1$$
Find $f(x) - g(x)$

A) $x^3 - x - 2$ B) $2x^2 + 4x - 5$

C) $-x^3 + x - 2$ D) $x^3 - x + 2$

118) Evaluate the below function.

$$g(x) = x^2 + 5x$$
$$h(x) = 4x - 2$$
Find $(g - h)(x)$

A) $-x^2 - x - 2$ B) $-x^2 + x - 2$

C) $x^2 + x + 2$ D) $6x + 1$

119) Evaluate the below function.

$$g(x) = x - 4$$
$$h(x) = x - 5$$
Find $g(x) - h(x)$

A) 1   B) -1

C) $2x^3 + 3x - 5$ D) $2x - 5$

120) Evaluate the below function.

$$h(x) = x^2 - 3$$
$$g(x) = x - 5$$
Find $h(x) - g(x)$

A) $-x^3 - 5x + 2$ B) $-2x^2 + 8x - 2$

C) $x^2 - x + 2$ D) $-x^2 + x - 2$

121) Evaluate the below function.

$$g(x) = 3x + 5$$
$$h(x) = 2x - 1$$
Find $(g - h)(x)$

A) $x - 6$ B) $-x - 6$

C) $x^2 - x + 4$ D) $x + 6$

122) Evaluate the below function.

$$h(n) = 4n + 2$$
$$g(n) = 4n - 3$$
Find $h(n) - g(n)$

A) $-n^3 + n - 5$ B) -5

C) 5   D) $-3n + 4$

123) Evaluate the below function.

$$g(a) = 4a - 3$$
$$f(a) = -a - 1$$
Find $g(a^2) - f(a^2)$

A) $-5a^2 + 2$ B) $-5a^2 - 2$

C) $10a - 2$ D) $5a^2 - 2$

124) Evaluate the below function.

$$f(t) = 2t - 5$$
$$g(t) = -t^2 + 1$$
Find $(f + g)(-2t)$

A) $-4t^2 - 4t - 4$

B) $-t^2 + 6t - 12$

C) $\dfrac{4t - 16 - t^2}{4}$

D) $-t^4 + 2t^2 - 4$

                   www.math-knots.com | www.a4ace.com

125) Evaluate the below function.

$$f(a) = a + 5$$
$$g(a) = 2a + 2$$

Find $f(-3a) - g(-3a)$

A) $3a + 3$ B) $-a^2 + 3$

C) $-3a - 3$ D) $-3a + 3$

126) Evaluate the below function.

$$h(t) = 4t - 1$$
$$g(t) = -2t^2 - 4$$

Find $h(-3t) + g(-3t)$

A) $-18t^2 - 12t - 5$ B) $-2t^2 - 3$

C) $-2t^2 - 4t - 5$ D) $-2t^2 + 20t - 53$

127) Evaluate the below function.

$$h(a) = 2a + 1$$
$$g(a) = a^2 - 5$$

Find $h(a^2) + g(a^2)$

A) $a^4 + 2a^2 - 4$

B) $4a^2 + 4a - 4$

C) $a^4 - 2a^2 - 4$

D) $9a^2 + 6a - 4$

128) Evaluate the below function.

$$f(n) = n^2 + 4n$$
$$g(t) = n + 5$$

Find $(f - g)(2n)$

A) $4n^2 + 6n - 5$

B) $4n - 5$

C) $-4n^2 - 6n + 5$

D) $-4n^2 + 6n + 5$

129) Evaluate the below function.

$$g(x) = 3x - 5$$
$$f(x) = -x - 4$$

Find $(g + f)(3x)$

A) $6x - 9$ B) $-6x - 9$

C) $\dfrac{2x - 27}{3}$ D) $-2x - 15$

130) Evaluate the below function.

$$f(t) = -t^2 - 5t$$
$$g(t) = 4t$$

Find $f(-t) - g(-t)$

A) $t^2 + 9t$ B) $t^2 - 9t$

C) $-t^2 + 9t$ D) $-t^3 + t^2 + 4t - 1$

Algebra 1

131) Evaluate the below function.

$$f(a) = 2a^2 - 1$$
$$g(a) = -2a + 1$$

Find $f\left(\dfrac{a}{4}\right) + g\left(\dfrac{a}{4}\right)$

A) $2a^2 - 17a + 31$

B) $2a^2 - a - 5$

C) $2a^2 + 6a + 4$

D) $\dfrac{a^2 - 4a}{8}$

132) Evaluate the below function.

$$f(x) = 3x + 2$$
$$g(x) = x - 4$$

Find $(f - g)(x - 2)$

A) $8x + 6$ B) $2x + 2$

C) $2x - 10$ D) $-2x - 2$

133) Evaluate the below function.

$$g(a) = a^2 - 5$$
$$h(a) = 2a - 1$$

Find $g(a^2) - h(a^2)$

A) $-a^4 - 2a^2 + 4$

B) $-a^4 + 2a^2 + 4$

C) $a^2 + 6a + 4$

D) $a^4 - 2a^2 - 4$

134) Evaluate the below function.

$$g(a) = a + 3$$
$$h(a) = 4a + 4$$

Find $g(a^2) - h(a^2)$

A) $3a^2 + 1$ B) $-3a - 4$

C) $-3a^2 + 1$ D) $-3a^2 - 1$

135) Evaluate the below function.

$$g(x) = 3x - 5$$
$$h(x) = x^2 - 4x$$

Find $g(x^2) + h(x^2)$

A) $x^4 + x^2 - 5$

B) $x^4 - x^2 - 5$

C) $\dfrac{-20 - 2x + x^2}{4}$

D) $x^2 + x - 5$

136) Evaluate the below function.

$$h(a) = 2a - 4$$
$$g(a) = a^2 - 4a$$

Find $h\left(\dfrac{a}{3}\right) - g\left(\dfrac{a}{3}\right)$

A) $\dfrac{12a - 16 - a^2}{4}$ B) $\dfrac{a^2 - 18a + 36}{9}$

C) $\dfrac{-36 + 18a - a^2}{9}$ D) $\dfrac{a^2 + 18a + 36}{9}$

137) Evaluate the below function.

$$h(x) = 3x + 1$$
$$g(x) = x^2 - 4$$

Find $h(4x) + g(4x)$

A) $x^2 - 3x - 3$

B) $16x^2 - 12x - 3$

C) $x^4 + 3x^2 - 3$

D) $16x^2 + 12x - 3$

138) Evaluate the below function.

$$f(x) = 3x + 3$$
$$g(x) = 2x + 2$$

Find $f\left(\dfrac{x}{2}\right) + g\left(\dfrac{x}{2}\right)$

A) $\dfrac{x^2 - 9x + 36}{9}$ B) $\dfrac{-5x + 10}{2}$

C) $\dfrac{5x + 10}{2}$ D) $\dfrac{6x + x^2 + 16}{4}$

139) Evaluate the below function.

$$h(t) = 2t - 1$$
$$g(t) = 3t + 5$$

Find $(h + g)\left(\dfrac{x}{3}\right)$

A) $\dfrac{-5x + 12}{3}$ B) $-5x + 4$

C) $\dfrac{5x + 12}{3}$ D) $-15x + 4$

140) Evaluate the below function.

$$f(x) = 2x + 1$$
$$g(x) = x^2 + 1$$

Find $f\left(\dfrac{x}{4}\right) + g\left(\dfrac{x}{4}\right)$

A) $\dfrac{32 - 8x + x^2}{16}$

B) $\dfrac{32 + 8x + x^2}{16}$

C) $4x^2 + 4x + 2$

D) $16x^2 - 8x + 2$

141) Evaluate the below function.

$$g(t) = 2t + 2$$
$$h(t) = t + 4$$

Find $g(z + 2) - h(z + 2)$

A) $z + 4$ B) $-z$

C) $z - 5$ D) z

142) Evaluate the below function.

$$f(x) = x^3 + 2x$$
$$g(x) = 4x + 3$$

Find $f(x^2) - g(x^2)$

A) $-x^2$

B) $-x^6 + 2x^2 + 3$

C) $x^6 - 2x^2 + 3$

D) $x^6 - 2x^2 - 3$

www.math-knots.com | www.a4ace.com

143) Evaluate the below function.

$$f(x) = x - 2$$
$$g(x) = x^3 + 3x$$

Find $(f - g)(-3x)$

A) $27x^3 + 6x + 2$

B) $27x^3 + 6x - 2$

C) $-64x^3 - 8x - 2$

D) $-27x^3 - 6x + 2$

144) Below graph represents a relation. Find the domain and range and state if the relation is function or not.

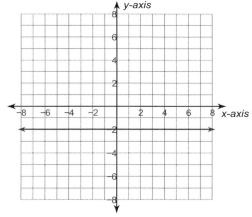

A) The relation is a function.
   Domain:All real numbers
   Range:All real numbers

B) The relation is a function.
   Domain:All real numbers
   Range:$y = -4$

C) The relation is a function.
   Domain:All real numbers
   Range:$y = -3$

D) The relation is a function.
   Domain:All real numbers
   Range:$y = -2$

145) Below graph represents a relation. Find the domain and range and state if the relation is function or not.

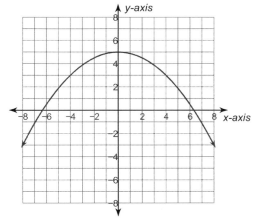

A) The relation is a function.
   Domain:All real numbers
   Range:$y \geq -4$

B) The relation is a function.
   Domain:All real numbers
   Range:$y \leq 5$

C) The relation is a function.
   Domain:All real numbers
   Range:$y \leq 2$

D) The relation is a function.
   Domain:All real numbers
   Range:$y \geq -3$

146) Below graph represents a relation. Find the domain and range and state if the relation is function or not.

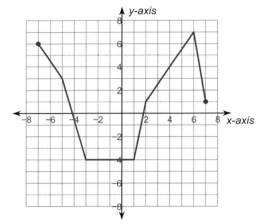

A) The relation is a function.
   Domain: $-7 \leq x \leq 7$
   Range: $-6 \leq y \leq 4$

B) The relation is a function.
   Domain: $-6 \leq x \leq 6$
   Range: $-7 \leq y \leq 3$

C) The relation is a function.
   Domain: $-6 \leq x \leq 6$
   Range: $-3 \leq y \leq 1$

D) The relation is a function.
   Domain: $-7 \leq x \leq 7$
   Range: $-4 \leq y \leq 7$

147) Below graph represents a relation. Find the domain and range and state if the relation is function or not.

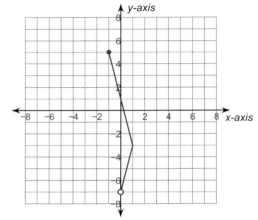

A) The relation is not a function.
   Domain: $-5 < x \leq 1$
   Range: $-4 < y < 5$

B) The relation is not a function.
   Domain: $-1 \leq x \leq 1$
   Range: $-7 < y \leq 5$

C) The relation is not a function.
   Domain: $-1 < x \leq 2$
   Range: $-6 \leq y < 6$

D) The relation is not a function.
   Domain: $-2 \leq x < 1$
   Range: $-4 < y \leq 6$

148) Below graph represents a relation.
Find the domain and range and
state if the relation is function or not.

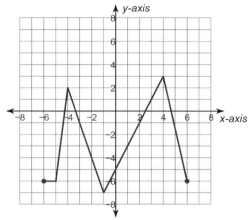

A) The relation is a function.
Domain: $-3 \le x \le 4$
Range: $-2 \le y \le 6$
B) The relation is a function.
Domain: $-6 \le x \le 6$
Range: $-7 \le y \le 3$
C) The relation is a function.
Domain: $-6 \le x \le 6$
Range: $-1 \le y \le 3$
D) The relation is a function.
Domain: $-6 \le x \le 6$
Range: $-7 \le y \le 5$

149) Below graph represents a relation.
Find the domain and range and
state if the relation is function or not.

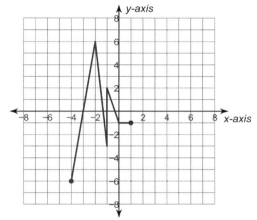

A) The relation is not a function.
Domain: $-4 \le x \le 1$
Range: $-6 \le y \le 6$
B) The relation is not a function.
Domain: $-7 \le x \le 7$
Range: $-7 \le y \le 5$
C) The relation is not a function.
Domain: $-7 \le x \le 5$
Range: $-4 \le y \le 5$
D) The relation is not a function.
Domain: $-3 \le x \le 7$
Range: $-5 \le y \le 6$

Algebra 1

150) Below graph represents a relation. Find the domain and range and state if the relation is function or not.

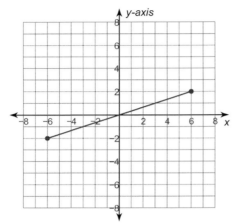

A) The relation is a function.
   Domain:$-5 < x \le 6$
   Range:$y = -6$
B) The relation is a function.
   Domain:$-6 < x \le 6$
   Range:$-3 \le y < 1$
C) The relation is a function.
   Domain:$-6 \le x \le 6$
   Range:$-2 \le y \le 2$
D) The relation is a function.
   Domain:$-6 < x < 6$
   Range:$-1 < y < 5$

151) Below graph represents a relation. Find the domain and range and state if the relation is function or not.

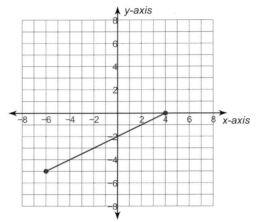

A) The relation is a function.
   Domain:$-4 < x \le 6$
   Range:$-3 \le y < 2$
B) The relation is a function.
   Domain:$-5 < x \le 4$
   Range:$-6 \le y < 3$
C) The relation is a function.
   Domain:$-6 < x < 5$
   Range:$-5 < y < 6$
D) The relation is a function.
   Domain:$-6 \le x \le 4$
   Range:$-5 \le y \le 0$

www.math-knots.com | www.a4ace.com

152) Below graph represents a relation.
Find the domain and range and
state if the relation is function or not.

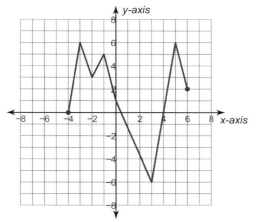

A) The relation is a function.
   Domain:$-6 \leq x \leq 6$
   Range:$-6 \leq y \leq 7$
B) The relation is a function.
   Domain:$-7 \leq x \leq 6$
   Range:$-7 \leq y \leq 5$
C) The relation is a function.
   Domain:$-7 \leq x \leq 6$
   Range:$-7 \leq y \leq 7$
D) The relation is a function.
   Domain:$-4 \leq x \leq 6$
   Range:$-6 \leq y \leq 6$

153) Below graph represents a relation.
Find the domain and range and
state if the relation is function or not.

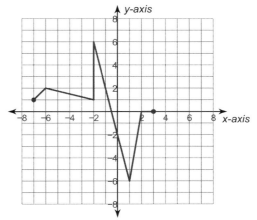

A) The relation is not a function.
   Domain:$-7 \leq x \leq 7$
   Range:$-1 \leq y \leq 5$
B) The relation is not a function.
   Domain:$-7 \leq x \leq 3$
   Range:$-6 \leq y \leq 6$
C) The relation is not a function.
   Domain:$-7 \leq x \leq 4$
   Range:$-6 \leq y \leq 5$
D) The relation is not a function.
   Domain:$-7 \leq x \leq 2$
   Range:$-6 \leq y \leq 4$

www.math-knots.com | www.a4ace.com

154) Below graph represents a relation. Find the domain and range and state if the relation is function or not.

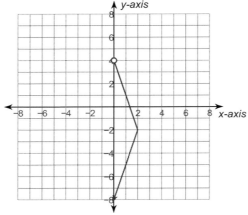

A) The relation is not a function.
   Domain: $x \leq 1$
   Range: $y < 4$
B) The relation is not a function.
   Domain: $x \leq 2$
   Range: $y < 4$
C) The relation is not a function.
   Domain: $x \geq -2$
   Range: $y < 6$
D) The relation is not a function.
   Domain: $x \leq 0$
   Range: $y > -4$

155) Below graph represents a relation. Find the domain and range and state if the relation is function or not.

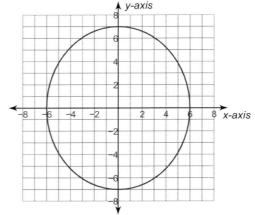

A) The relation is not a function.
   Domain: $-4 \leq x \leq 4$
   Range: $-3 \leq y \leq 5$
B) The relation is not a function.
   Domain: $-6 \leq x \leq 6$
   Range: $-7 \leq y \leq 7$
C) The relation is not a function.
   Domain: $-3 \leq x \leq 3$
   Range: $-2 \leq y \leq 6$
D) The relation is not a function.
   Domain: $-2 \leq x \leq 6$
   Range: $-3 \leq y \leq 3$

www.math-knots.com | www.a4ace.com

156) Below graph represents a relation. Find the domain and range and state if the relation is function or not.

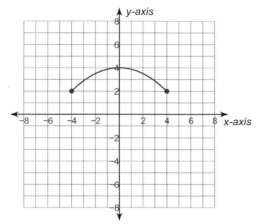

A) The relation is a function.
Domain:$[-7, 5)$
Range:$[-4, 4]$
B) The relation is a function.
Domain:$(-7, 5]$
Range:$[-3, 5)$
C) The relation is a function.
Domain:$[-4, 4]$
Range:$[2, 4]$
D) The relation is a function.
Domain:$(-7, 5]$
Range:$(-3, 5]$

157) Below graph represents a relation. Find the domain and range and state if the relation is function or not.

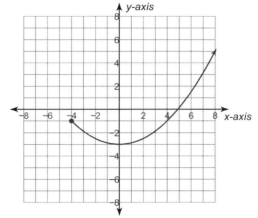

A) The relation is a function.
Domain:$(-\infty, 7)$
Range:$(-\infty, 6]$
B) The relation is a function.
Domain:$[-4, \infty)$
Range:$[-3, \infty)$
C) The relation is a function.
Domain:$(-\infty, 5)$
Range:$[-4, \infty)$
D) The relation is a function.
Domain:$[-7, \infty)$
Range:$(-\infty, 2]$

158) Below graph represents a relation. Find the domain and range and state if the relation is function or not.

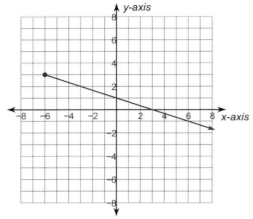

A) The relation is a function.
   Domain: $(-\infty, 4)$
   Range: $(-2, \infty)$
B) The relation is a function.
   Domain: $[-6, \infty)$
   Range: $(-\infty, 3]$
C) The relation is a function.
   Domain: $(-\infty, 6]$
   Range: $[-2, \infty)$
D) The relation is a function.
   Domain: $[-5, \infty)$
   Range: $[-7, \infty)$

159) Below graph represents a relation. Find the domain and range and state if the relation is function or not.

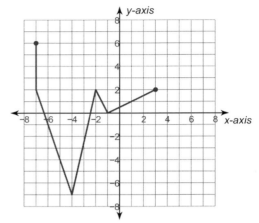

A) The relation is not a function.
   Domain: $[-2, 7]$
   Range: $[-6, 4]$
B) The relation is not a function.
   Domain: $[-2, 7]$
   Range: $[-6, 7]$
C) The relation is not a function.
   Domain: $[-7, 5]$
   Range: $[-5, 5]$
D) The relation is not a function.
   Domain: $[-7, 3]$
   Range: $[-7, 6]$

www.math-knots.com | www.a4ace.com

160) Below graph represents a relation. Find the domain and range and state if the relation is function or not.

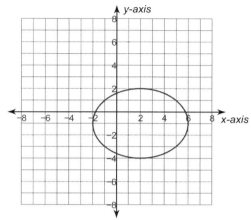

A) The relation is not a function.
   Domain:$[-4, 6]$
   Range:$[-3, 5]$
B) The relation is not a function.
   Domain:$[-7, 7]$
   Range:$[-5, 7]$
C) The relation is not a function.
   Domain:$[-7, 3]$
   Range:$[-7, 7]$
D) The relation is not a function.
   Domain:$[-2, 6]$
   Range:$[-4, 2]$

161) Below graph represents a relation. Find the domain and range and state if the relation is function or not.

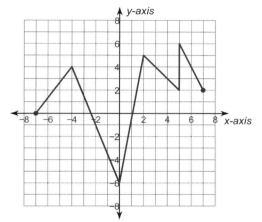

A) The relation is not a function.
   Domain:$[-7, 7]$
   Range:$[-6, 6]$
B) The relation is not a function.
   Domain:$[-5, 6]$
   Range:$[-7, 7]$
C) The relation is not a function.
   Domain:$[-7, 7]$
   Range:$[-6, 7]$
D) The relation is not a function.
   Domain:$[-6, 7]$
   Range:$[-6, 6]$

www.math-knots.com | www.a4ace.com

162) Below graph represents a relation. Find the domain and range and state if the relation is function or not.

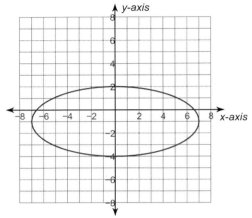

A) The relation is not a function.
   Domain:$[-7, 7]$
   Range$[-4, 2]$
B) The relation is not a function.
   Domain:$[-2, 4]$
   Range$[-4, 2]$
C) The relation is not a function.
   Domain:$[-7, 3]$
   Range$[-6, 2]$
D) The relation is not a function.
   Domain:$[-2, 4]$
   Range$[-5, 1]$

163) Below graph represents a relation. Find the domain and range and state if the relation is function or not.

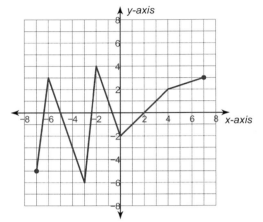

A) The relation is a function.
   Domain:$[-6, 5]$
   Range$[-7, 6]$
B) The relation is a function.
   Domain:$[-7, 3]$
   Range$[-7, 3]$
C) The relation is a function.
   Domain:$[-6, 5]$
   Range$[-2, 6]$
D) The relation is a function.
   Domain:$[-7, 7]$
   Range$[-6, 4]$

www.math-knots.com | www.a4ace.com

164) Below graph represents a relation. Find the domain and range and state if the relation is function or not.

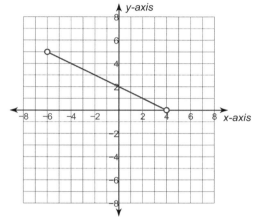

A) The relation is a function.
  Domain:$(-7, 5)$
  Range$\{3\}$
B) The relation is a function.
  Domain:$(-6, 6]$
  Range$\{-5\}$
C) The relation is a function.
  Domain:$[-4, 6]$
  Range$[-4, 6]$
D) The relation is a function.
  Domain:$(-6, 4)$
  Range:$(0, 5)$

165) Below graph represents a relation. Find the domain and range and state if the relation is function or not.

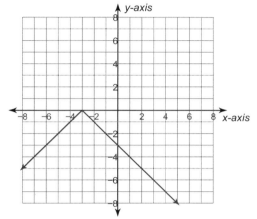

A) The relation is a function.
  Domain:$(-\infty, \infty)$
  Range$[-2, \infty)$
B) The relation is a function.
  Domain:$(-\infty, \infty)$
  Range$(-\infty, 0]$
C) The relation is a function.
  Domain:$(-\infty, \infty)$
  Range$(-\infty, 1]$
D) The relation is a function.
  Domain:$(-\infty, \infty)$
  Range$[-1, \infty)$

www.math-knots.com | www.a4ace.com

166) $f(x) = x^2 + 1$

$g(x) = -3x + 4$

Find $(f - 3g)(-8)$

A) −19   B) 41

C) −167   D) −11

167) $g(t) = t^2 - 2t$

$f(t) = t - 1$

Find $(g \circ f)(-7)$

A) 3   B) 34

C) 62   D) 80

168) $h(t) = 2t + 4$

$g(t) = t + 5$

Find $(-5h + 4g)(-5)$

A) 0   B) 6

C) 30   D) −24

169) $g(n) = n - 5$

$h(n) = n^2 - 4n$

Find $g(h(-4))$

A) 27   B) −8

C) 117   D) 5

170) $f(t) = 4t + 1$

$g(t) = 2t - 2$

Find $(-3f + 3g)(2)$

A) −3   B) −21

C) 21   D) −39

171) $g(n) = 3n + 3$

$f(n) = 3n - 3$

Find $g(-2) \cdot f(-2)$

A) 0   B) −25

C) 27   D) 135

172) $g(t) = t^2 - 5$

$f(t) = -t - 4$

Find $(g \circ f)(8)$

A) 193   B) 11

C) 139   D) −63

173) $g(x) = -3x + 3$

$f(x) = -3x^3 - 3x^2$

Find $g(2) \cdot f(2)$

A) 0   B) −42

C) −2   D) 108

174) $h(n) = 3n + 4$

$g(n) = 3n - 1$

Find $(h \circ g)(-10)$

A) −89   B) 22

C) −140   D) −79

175) $h(n) = 3n^2 - 4 + 2n$

$g(n) = 2n - 2$

Find $(h - 5g)(3)$

A) −93   B) −141

C) −62   D) 9

176) $h(x) = 3x^2 - 4x$

   $g(x) = 3x + 4$

   Find $(2h - 4g)(7)$

   A) -22   B) 48

   C) -32   D) 138

177) $h(x) = x^2 - x$

   $g(x) = x - 1$

   Find $(h \circ g)(-5)$

   A) 19   B) 42

   C) 72   D) 29

178) $g(a) = 3a + 1$

   $f(a) = 2a + 3$

   Find $g(-6) + 5f(-6)$

   A) -94   B) 110

   C) 120   D) -62

179) $g(x) = 4x + 1$

   $h(x) = x^2 - 5x$

   Find $(g \cdot h)(2)$

   A) -56   B) -54

   C) -18   D) -80

180) Given $g(x) = 3x$; Find x if g(x) = 27

181) $g(x) = -x^2 - 3x$

   $h(x) = 2x + 3$

   Find $g(-5) \cdot h(-5)$

   A) 70   B) 2

   C) 27   D) -2

182) $g(x) = 3x + 2$

   $f(x) = -4x + 5$

   Find $(g \circ f)(-5)$

   A) 77   B) -63

   C) 57   D) 41

183) $f(n) = n + 3$

   $g(n) = n + 5$

   Find $(f \cdot g)(8)$

   A) 0   B) 143

   C) 120   D) 15

184) $g(a) = -3a + 1$

   $f(a) = -4a + 4$

   Find $(g \cdot f)(2)$

   A) 20   B) 4

   C) 160   D) 84

185) Given $h(t) = -2t - 3$;
   Find t if h(t) = -15

186) Given $g(n) = 3n - 5$; Find n if g(n) = -2

187) Given $k(a) = 2a + 5$; Find a if k(a) = 15

188) Given $h(x) = x + 5$; Find x if h(x) = -4

189) Given $h(x) = x - 2$; Find x if h(x) = -3

190) Given $g(t) = t + 3$; Find t if g(t) = 1

191) Given $g(x) = 2x + 3$;
Find x if g(x) = 15

192) Given $g(x) = 3x - 4$;
Find x if g(x) = 8

193) Given $k(x) = 2x - 5$;
Find x if k(x) = -1

194) Given $g(a) = 2a + 5$;
Find a if g(a) = 11

195) Given $h(a) = 3a$;
Find a if h(a) = 15

196) Given $g(a) = 3a + 4$; Find a if g(a) = 13

197) Find the 32nd term in the following series

6, 16, 26, 36, ...

198) Find the 36th term in the following series

35, 29, 23, 17, ...

199) Find the 28th term in the following series

5, 15, 25, 35, ...

200) Find the 32nd term in the following series

1, 9, 17, 25, ...

201) Find the 32nd term in the following series

−23, −123, −223, −323, ...

202) Find the 29th term in the following series

19, 49, 79, 109, ...

203) Find the 34th term in the following series

21, 26, 31, 36, ...

204) Find the 24th term in the following series

3, −6, −15, −24, ...

205) Find the 40th term in the following series

−14, −34, −54, −74, ...

206) Find the next two terms in the below sequence of numbers

37, 57, 77, 97, ...

207) Find the next two terms in the below sequence of numbers

−4, −14, −24, −34, ...

208) Find the next two terms in the below sequence of numbers

−5, 195, 395, 595, ...

209) Find the next two terms in the below sequence of numbers

−7, −16, −25, −34, ...

210) Find the next two terms in the below sequence of numbers

14, 8, 2, −4, ...

211) Find the next two terms in the below sequence of numbers

39, 47, 55, 63, ...

212) Find the next two terms in the below sequence of numbers

18, 22, 26, 30, ...

1) Evaluate the below function.

$h(a) = -3|2a - 3| + 1$; Find $h(6)$

A) −68   B) −8

C) −38   D) −26

2) Solve the inequality and graph its solution.

$\left|-6 + x\right| \geq 3$

3) For which of the following values of x is this inequality is true ?

$16x + 15 \geq -609$

A) $x \geq -39$ B) $x \geq -49$

C) $x \leq -17$ D) $x \geq -17$

4) For which of the following values of b is this inequality is true ?

$b - 48 > -78$

A) $b > -30$ B) $b > 30$

C) $b > -\dfrac{13}{8}$ D) $b > 3744$

5) Evaluate the below function.

$f(a) = 3a + 2$

$g(a) = a^2 - 3a$

Find $(f - g)(-4)$

A) 38   B) − 38

C) − 10   D) − 89

6) For which of the following values of n is this inequality is true ?

$-20n + 7 \leq 407$

A) $n \geq -20$ B) $n \leq -64$

C) $n \leq -10$ D) $n \geq -64$

7) For which of the following values of v is this inequality is true ?

$-22 < v - (-25)$

A) $v < 3$ B) $v > 3$

C) $v > -47$ D) $v > \dfrac{22}{25}$

8) For which of the following values of x is this inequality is true ?

$x + 22 \leq 7$

A) $x \leq -29$ B) $x \geq -15$

C) $x \leq -15$ D) $x \leq 29$

9) Solve the inequality and graph its solution.

$$-7 + \left|-3x + 3\right| > 17$$

10) Evaluate the below function.

$$g(n) = 2\left|n + 1\right| + 1; \text{ Find } g(-2 - n)$$

A) $2\left|\dfrac{3 + n}{3}\right| + 1$

B) $2n^2 + 3$

C) $2\left|-4n + 1\right| + 1$

D) $2\left|-1 - n\right| + 1$

11) Solve the inequality and graph its solution.

$$-3 + \left|v - 10\right| \geq -2$$

12) For which of the following values of n is this inequality and the graph is true ?

$$-87.72 > 8.6n$$

A) $n < -79.12$:
-86 -84 -82 -80 -78 -76

B) $n < 79.12$:
78 80 82 84 86

C) $n < -10.2$:
-12 -10 -8 -6 -4

D) $n < -79.12$:
-84 -82 -80 -78 -76 -74

13) For which of the following values of v is this inequality and the graph is true ?

$$\dfrac{v}{12.1} < -2.9$$

A) $v < -35.09$:
-38 -36 -34 -32 -30

B) $v > -9.2$:
-10 -8 -6 -4 -2

C) $v > -9.2$:
-10 -8 -6 -4 -2

D) $v < -35.09$:
-42 -40 -38 -36 -34 -32

14) Evaluate the below function.

$$g(n) = 3n - 3$$
$$h(n) = 3n^2 - 4n$$
Find $g(4) - h(4)$

A) 1   B) 23

C) – 23   D) 79

15) For which of the following values of v is this inequality is true ?

$$-47v < -329$$

A) $v > 15463$ B) $v < 15463$

C) $v > 7$ D) $v < -376$

16) Solve the inequality and graph its solution.

$$10 + 6\left|10x - 4\right| \geq 46$$

17) Solve the compound inequality and graph its solution. Represent the solution in :
Set builder notation and interval notation .

$$29 \leq 10n - 1 \leq 49$$

A) $n \geq -14$:

B) $3 \leq n \leq 5$:

C) $-14 \leq n < 4$:

D) $n < 4$:

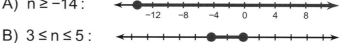

18) For which of the following values of r is this inequality is true ?

$$-\frac{32}{19} < \frac{r}{19}$$

A) $r > -32$ B) $r < -32$

C) $r < \frac{32}{361}$ D) $r > \frac{32}{361}$

19) For which of the following values of x is this inequality and the graph is true ?

$$-0.45 \leq x + (-2)$$

A) $x \geq -0.9$:

B) $x \geq 1.55$:

C) $x \geq 1.55$:

D) $x \leq 1.55$:

20) Solve the inequality and graph its solution.

$$4\left|-7x + 2\right| - 6 \leq 98$$

21) For which of the following values of x is this inequality is true ?

$$11 > \frac{x + 7}{3}$$

A) $x > -47$ B) $x > 26$

C) $x > -59$ D) $x < 26$

www.math-knots.com | www.a4ace.com

22) Solve the compound inequality and graph its solution. Represent the solution in :
Set builder notation and interval notation .

$$-14 \le -2r - 8 \le 10$$

A) $r \le 3$:

B) $r \ge -9$:

C) No solution.:

D) $-9 \le r \le 3$:

23) Evaluate the below function.

$$h (x) = | x - 2 | ; \text{ Find } h (4)$$

A) 2 B) 10

C) 4 D) 8

24) Evaluate the below function.

$$g(n) = -4^{n + 1} - 2; \text{ Find } g(-2)$$

A) -66 B) -3

C) $-\dfrac{9}{4}$ D) -6

25) Solve the inequality and graph its solution.

$$-6 + \left| \dfrac{r}{3} \right| \ge -3$$

26) Evaluate the below function.

$$g (a) = 4 a + 3$$
$$h (a) = a^{3} - 3$$

Find $g (a) - h (a)$

A) $a^{3} - 4a - 6$ B) $-a^{3} + 4a + 6$

C) $a^{2} - 3a - 3$ D) $-a^{3} + 4a - 6$

27) Evaluate the below function.

$$f (x) = 4 x - 2$$
$$g (x) = -x + 1$$

Find $(f - g) (9)$

A) 48 B) -42

C) 22 D) 42

28) Evaluate the below function.

$$h(x) = 4^{x} ; \text{ Find } h(3x)$$

A) $\dfrac{1}{4^{x}}$ B) $\dfrac{1}{4^{3x}}$

C) 4^{3x} D) $4^{x^{2}}$

Algebra 1

29) Solve the compound inequality and graph its solution. Represent the solution in :
Set builder notation and interval notation .

$12m - 8 \geq -32 \text{ or } 2m + 8 \leq -2$

A) No solution.:
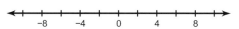

B) $m \leq -5$:
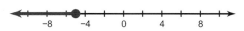

C) $m > 7$:

D) $m \geq -2 \text{ or } m \leq -5$:

30) Solve the compound inequality and graph its solution. Represent the solution in :
Set builder notation and interval notation .

$-6.7 \leq x + 0.2 \leq 3.3$

A) $-6.9 \leq x \leq 3.1$:

B) { All real numbers. }:

C) $3.3 < x < 4.1$:

D) $x < 4.1$:

31) Solve the compound inequality and graph its solution. Represent the solution in :
Set builder notation and interval notation .

$x - 6.1 < -9.6 \text{ and} -1.5 + x > -11.4$

A) $x > 4.3$:
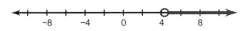

B) $-8.14 \leq x \leq 7.7$:
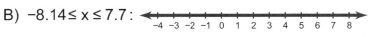

C) $-9.9 < x < -3.5$:

D) $x \geq -3.3$:

www.math-knots.com | www.a4ace.com

32)   Solve the compound inequality and graph its solution. Represent the solution in :
Set builder notation and interval notation .

$$b + \frac{17}{6} \geq \frac{13}{3} \text{ or } b + \frac{7}{10} < \frac{11}{10}$$

A) $b < -7$:

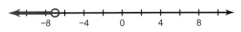

B) $b \geq \frac{3}{2} \text{ or } b < \frac{2}{5}$:

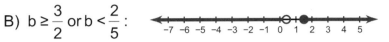

C) $b \geq 6$:

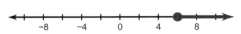

D) $b > 7$:

33)   Solve the compound inequality and graph its solution. Represent the solution in :
Set builder notation and interval notation .

$$x + 1 \geq \frac{4}{5} \text{ and } x + \frac{17}{4} < \frac{117}{20}$$

A) $-\frac{1}{5} \leq x < \frac{8}{5}$:

B) $x < \frac{8}{5}$:

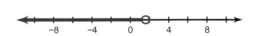

C) $-\frac{2}{7} < x \leq \frac{6}{5}$:

D) $\frac{21}{10} \leq x \leq \frac{51}{10}$:

34)   Solve the compound inequality and graph its solution. Represent the solution in :
Set builder notation and interval notation .

$$-\frac{9}{5} \geq k - \frac{3}{10} \geq -\frac{23}{10}$$

A) $-2 \leq k \leq -\frac{3}{2}$:

B) $-\frac{17}{9} < k \leq -\frac{10}{9}$:

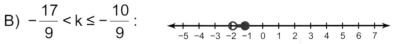

C) $\frac{5}{4} < k < \frac{14}{5}$:

D) $-\frac{3}{2} < k \leq 2$:

             www.math-knots.com | www.a4ace.com

35) Which graph represents the linear inequality and its solution.

$$x + y \geq 2$$

A)

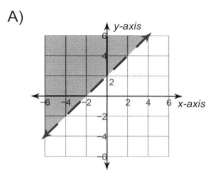

B)

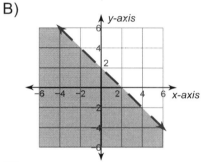

C)

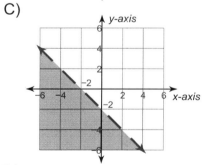

D)

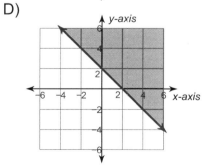

36) For which of the following values of x is this inequality is true ?

$$-7 > \frac{x}{7} - 2$$

A) x > − 35   B) x > − 60

C) x < − 35   D) x > − 93

37) For which of the following values of x is this inequality and the graph is true ?

$$9.09 > 12.19 + x$$

A) x > 21.28 :
B) x > −3.1 :
C) x < −3.1 :
D) x > −3.1 :

38) Solve the inequality and graph its solution.

$$7\left|4 + 7r\right| - 7 < 70$$

39) For which of the following values of k is this inequality is true ?

$$\frac{k}{37} < 2$$

A) k > −39   B) k < −74

C) k < −39   D) k < 74

www.math-knots.com | www.a4ace.com

Algebra 1

40) For which of the following values of v is this inequality is true ?

$$41 + v < 81$$

A) $v > 40$　　B) $v > \dfrac{81}{41}$

C) $v < 40$　　D) $v > 122$

41) Solve the inequality and graph its solution.

$$\dfrac{|a-9|}{2} < 4$$

42) For which of the following values of x is this inequality is true ?

$$1276 < 44x$$

A) $x < 29$　　B) $x > 29$

C) $x < 1320$　　D) $x < -29$

43) Evaluate the below function.

$$h(n) = n^3 + 5n^2; \text{ Find } h(3)$$

A) 16　　B) -192

C) 12　　D) 72

44) Which graph represents the linear inequality and its solution.

$$8x - 3y < -12$$

A)
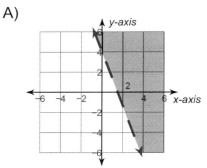

B)
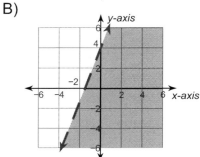

C)
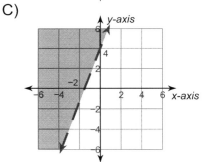

D)
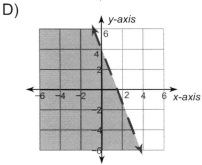

www.math-knots.com | www.a4ace.com

45) Which graph represents the linear inequality and its solution.

$$x - 4y > -4$$

A)

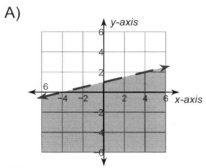

B)

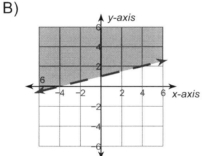

C)

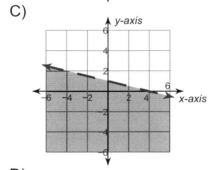

D)

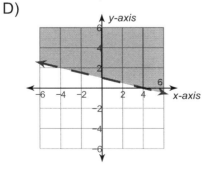

46) Solve the inequality and graph its solution.

$$-|4n - 10| + 3 \geq 1$$

47) Evaluate the below function.

$$p(x) = 3x - 1 ; \text{ Find } p(6)$$

A) -25 B) -4

C) 29 D) 17

48) Evaluate the below function.

$$p(x) = 2x ; \text{ Find } p(-9)$$

A) -18 B) 14

C) -12 D) -8

49) Solve the inequality and graph its solution.

$$10 + |8n + 1| \leq 43$$

www.math-knots.com | www.a4ace.com

50) Solve the inequality and graph its solution.

$$|p - 9| \geq 17$$

51) Evaluate the below function.

$$h(t) = 2 \cdot 3^{2\,t}; \text{ Find } h\left(\frac{t}{3}\right)$$

A) $2 \cdot 3^{2t-2}$ B) $2 \cdot 3^{2t-8}$

C) $2 \cdot 3^{\frac{1}{2}t}$ D) $2 \cdot 3^{\frac{2}{3}t}$

52) Solve the inequality and graph its solution.

$$|b + 2| \geq 1$$

53) Solve the inequality and graph its solution.

$$\left|\frac{r}{10}\right| < 4$$

54) Which graph represents the linear inequality and its solution.

$$x > 5$$

A)
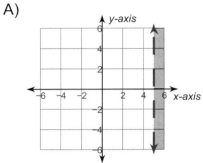

B)
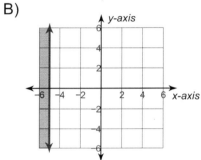

C)
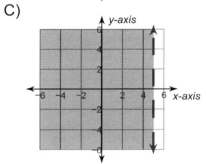

D)
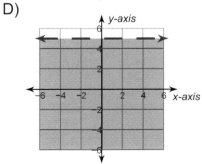

www.math-knots.com | www.a4ace.com

55) Solve the inequality and graph its solution.

$$|4x| + 6 \le 10$$

56) Evaluate the below function.

$$g(x) = 3|x - 3|; \text{ Find } g(x + 2)$$

A) $3|x - 1|$ B) $3|x^2 - 3|$

C) $3|-x - 3|$ D) $3|4x - 3|$

57) Evaluate the below function.

$$h(x) = 5^{-x-2}; \text{ Find } h(0)$$

A) $\dfrac{1}{25}$ B) $\dfrac{1}{5}$

C) $\dfrac{1}{125}$ D) 1

58) Evaluate the below function.

$$p(x) = -2 \cdot 2^{x-2} + 1; \text{ Find } p(-4t)$$

A) $-2^{t+1} + 1$ B) $-2^{-4t-1} + 1$

C) $-2^{t+2} + 1$ D) $-2^{2t-1} + 1$

59) Which graph represents the linear inequality and its solution.

$$y \ge 1$$

A)
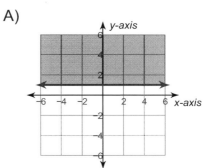

B)
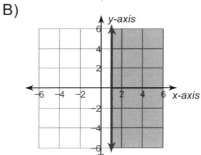

C)
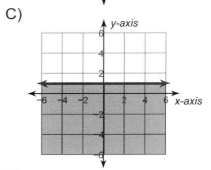

D)
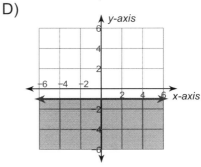

www.math-knots.com | www.a4ace.com

60) Which graph represents the linear inequality and its solution.

$$x > 1$$

A)

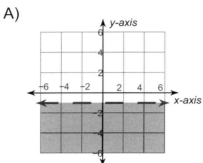

B)

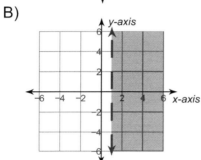

C)

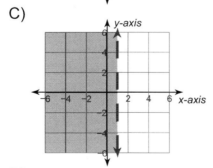

D)

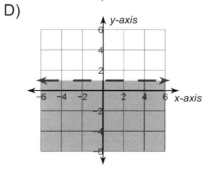

61) Evaluate the below function.

$$g(x) = 4x + 1$$

Given : $\{-7, -3, -2, 1, 2, 7\}$

Domain :

Range :

62) Evaluate the below function.

$$g(x) = 2|-3x| \; ; \; \text{Find } g(2)$$

A) 12   B) 54

C) 36   D) 6

63) Evaluate the below function.

$$g(n) = -5^{-n+1} - 1; \; \text{Find } g(3+n)$$

A) $-5^{3n+1} - 1$ B) $-5^{-3n+1} - 1$

C) $-5^{-2-n} - 1$ D) $-5^{4n+1} - 1$

64) Solve the inequality and graph its solution.

$$9|7r - 9| + 9 > 27$$

Algebra 1

65) Solve the compound inequality and graph its solution. Represent the solution in :
Set builder notation and interval notation .

$7x - 11 < 3$ or $11 + 6x \geq 83$

A) $x < 2$ or $x \geq 12$:

B) $x < -8$:

C) $x < -7$:

D) $x < 2$:

66) Solve the compound inequality and graph its solution. Represent the solution in :
Set builder notation and interval notation .

$-73 \leq -14m - 3 < 67$

A) $m \leq 5$: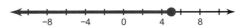

B) $-5 < m \leq 5$: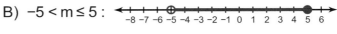

C) $m > 4$: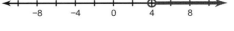

D) $m < 11$:

67) Evaluate the below function.

$h(a) = a^2 + 4a$

Given : $\{-5, -3, -2, 1, 2, 5\}$

Domain :

Range :

68) Evaluate the below function.

$g(x) = x^2 + 2$

$h(x) = 2x + 3$

Find $g(-5) - h(-5)$

A) 34   B) -34

C) 14   D) -14

www.math-knots.com | www.a4ace.com

69) Evaluate the below function.

$$f(x) = x^2 - 2 - x$$
$$g(x) = 2x + 1$$

Find $f(x) - g(x)$

A) $-x^2 - 3x + 3$ B) $x^2 - 3x - 3$

C) $6x - 2$ D) $-x^2 + 3x + 3$

70) Solve the inequality and graph its solution.

$$1 + |9 - 6m| \geq 34$$

71) Evaluate the below function.

$$h(x) = x^2 - 2x$$
$$g(x) = 2x + 4$$

Find $(h - g)(x)$

A) $6x - 3$ B) $x^2 - 4x - 4$

C) $-x^2 - 4x + 4$ D) $-x^2 + 4x + 4$

72) Evaluate the below function.

$$f(x) = x^3 - 3x$$
$$g(x) = 4x - 3$$

Find $(f - g)(x)$

A) $x^3 - 7x + 3$

B) $-x - 7$

C) $-x^3 + 7x - 3$

D) $x^3 - 7x - 3$

73) Evaluate the below function.

$$g(t) = 4t + 5$$
$$h(t) = 2t - 1$$

Find $g(t^2) + h(t^2)$

A) $-6t^2 + 4$ B) $6t^2 + 4$

C) $24t + 4$ D) $18t + 4$

74) Solve the inequality and graph its solution.

$$2 + 5|v + 4| \leq 72$$

75) Evaluate the below function.

$$g(n) = 3n + 1$$
$$f(n) = n^3 + 5n^2$$

Find $g(-3n) - f(-3n)$

A) $27n^3 - 45n^2 - 9n + 1$

B) $27n^3 + 45n^2 - 9n - 1$

C) $-27n^3 + 45n^2 + 9n - 1$

D) $-n^3 - 14n^2 - 54n - 62$

76) Evaluate the below function.

$$h(n) = n^2 - 3n$$

Given : $\{-6, -5, -1, 7, 10\}$

Domain :

Range :

www.math-knots.com | www.a4ace.com

77) Evaluate the below function.

$$h(x) = 2x + 5$$
$$g(x) = x^2 + 2x$$

Find $h(x + 3) + g(x + 3)$

A) $x^4 + 4x^2 + 5$

B) $x^2 + 10x + 26$

C) $x^2 + 2x + 2$

D) $\dfrac{45 + 12x + x^2}{9}$

78) Evaluate the below function.

$$g(n) = 3n + 2$$
$$h(n) = -3n^3 + 1$$

Find $(g + h)\left(\dfrac{n}{3}\right)$

A) $-24n^3 + 6n + 3$

B) $\dfrac{-9n + 27 + n^3}{9}$

C) $\dfrac{9n + 27 - n^3}{9}$

D) $24n^3 - 6n + 3$

79) Find the next two terms in the below sequence of numbers

−4, 26, 56, 86, ...

80) Find the 23rd term in the following series

19, 15, 11, 7, ...

81) $f(x) = x^3 + 5x$

$g(x) = x + 1$

Find $f(g(-4))$

A) 85   B) −42

C) −83   D) −18

82) Given $h(t) = 2t$; Find t if h(t) = -14

83) Given $w(t) = -3t - 3$; Find t if w(t) = 24

84) Find the next two terms in the below sequence of numbers

11, 211, 411, 611, ...

85) $g(x) = 2x$

$h(x) = 3x^3 + 5x$

Find $(g \cdot h)(-2)$

A) 72   B) 0
C) 136   D) 24

86) Find the next two terms in the below sequence of numbers

-20, -25, -30, -35, ...

1) Find the slope of the straight line passing through the points

$$A(-12, 10), B(-13, 8)$$

A) $-\dfrac{1}{2}$ B) $\dfrac{1}{2}$

C) 2   D) -2

2) Find the slope of the straight line passing through the points

$$P(3, 9), Q(-9, -9)$$

A) $-\dfrac{3}{2}$ B) $-\dfrac{2}{3}$

C) $\dfrac{2}{3}$ D) $\dfrac{3}{2}$

3) Find the slope of the straight line passing through the points

$$R(10, -7), S(0, 3)$$

A) -2 B) -1

C) 2   D) 1

4) Find the slope of the straight line passing through the points

$$M(6, -20), N(6, 15)$$

A) $-\dfrac{1}{5}$ B) $\dfrac{1}{5}$

C) 0   D) Undefined

5) Find the slope of the straight line passing through the points

$$E(-6, -2), F(-8, -17)$$

A) $-\dfrac{2}{15}$ B) $-\dfrac{15}{2}$

C) $\dfrac{2}{15}$ D) $\dfrac{15}{2}$

6) Find the slope of the straight line passing through the points

$$S(-19, -17), T(10, 8)$$

A) $\dfrac{29}{25}$ B) $-\dfrac{29}{25}$

C) $\dfrac{25}{29}$ D) $-\dfrac{25}{29}$

7) Find the slope of the straight line passing through the points

$$P(7, -6), Q(-17, 0)$$

A) -4 B) $-\dfrac{1}{4}$

C) $\dfrac{1}{4}$ D) 4

8) Find the slope of the straight line passing through the points

$$A(-8, 13), B(-13, 10)$$

A) $\dfrac{3}{5}$ B) $\dfrac{5}{3}$

C) $-\dfrac{5}{3}$ D) $-\dfrac{3}{5}$

9) Find the slope of the straight line passing through the points

$$P(11, 8), Q(15, 20)$$

A) -3 B) $\dfrac{1}{3}$

C) $-\dfrac{1}{3}$ D) 3

10) Find the slope of the straight line passing through the points

$$R(2, 5), S(9, 3)$$

A) $\dfrac{7}{2}$ B) $-\dfrac{7}{2}$

C) $-\dfrac{2}{7}$ D) $\dfrac{2}{7}$

11) Find the slope of the straight line passing through the points

$$M(18, 18), N(6, 4)$$

A) $-\dfrac{6}{7}$ B) $\dfrac{7}{6}$

C) $-\dfrac{7}{6}$ D) $\dfrac{6}{7}$

12) Find the slope of the straight line passing through the points

$$A(13, 16), B(13, 19)$$

A) Undefined   B) $\dfrac{3}{4}$

C) 0 D) $-\dfrac{3}{4}$

13) Find the slope of the straight line passing through the points

$$S(4, 2), T(-2, -1)$$

A) -2 B) $\dfrac{1}{2}$

C) 2 D) $-\dfrac{1}{2}$

14) Find the slope of the straight line passing through the points

$$A(-7, 0), B(3, 9)$$

A) $-\dfrac{9}{10}$ B) $-\dfrac{10}{9}$

C) $\dfrac{10}{9}$ D) $\dfrac{9}{10}$

15) Find the slope of the straight line passing through the points

$$E(7, -1), F(-20, -2)$$

A) 27 B) $-\dfrac{1}{27}$

C) -27 D) $\dfrac{1}{27}$

16) Find the slope of the straight line passing through the points

$$M(7, -8), N(-11, -1)$$

A) $\dfrac{18}{7}$ B) $-\dfrac{18}{7}$

C) $-\dfrac{7}{18}$ D) $\dfrac{7}{18}$

 **Algebra 1**

17) Find the slope of the straight line passing through the points

$$R(-2,1), S(-12,6)$$

A) $\dfrac{1}{2}$ B) $-\dfrac{1}{2}$

C) 2   D) -2

18) Find the slope of the straight line passing through the points

$$A(-8,7), B(16,-8)$$

A) $-\dfrac{8}{5}$ B) $\dfrac{8}{5}$

C) $\dfrac{5}{8}$ D) $-\dfrac{5}{8}$

19) Find the slope of the straight line passing through the points

$$P(7,-1), Q(-15,10)$$

A) $\dfrac{1}{2}$ B) 2

C) -2 D) $-\dfrac{1}{2}$

20) Find the slope of the straight line passing through the points

$$R(19,-2), S(-17,13)$$

A) $-\dfrac{5}{12}$ B) $\dfrac{5}{12}$

C) $-\dfrac{12}{5}$ D) $\dfrac{12}{5}$

21) Find the slope of the straight line passing through the points

$$M(-13,6), N(17,-14)$$

A) $-\dfrac{3}{2}$ B) $-\dfrac{2}{3}$

C) $\dfrac{2}{3}$ D) $\dfrac{3}{2}$

22) Find the slope of the straight line passing through the points

$$E(-7,-13), F(1,3)$$

A) -2 B) $-\dfrac{1}{2}$

C) 2   D) $\dfrac{1}{2}$

23) Find the slope of the straight line passing through the points

$$A(4,16), B(2,10)$$

A) $-\dfrac{1}{3}$ B) 3

C) -3 D) $\dfrac{1}{3}$

24) Find the slope of the straight line passing through the points

$$P(-11,14), Q(4,7)$$

A) $\dfrac{15}{7}$ B) $\dfrac{7}{15}$

C) $-\dfrac{7}{15}$ D) $-\dfrac{15}{7}$

www.math-knots.com | www.a4ace.com

25) Find the slope of the straight line passing through the points

$$R(-17,-18), S(-17,13)$$

A) Undefined   B) 0

C) $-\dfrac{3}{2}$ D) $\dfrac{3}{2}$

26) Find the slope of the straight line passing through the points

$$M(3,13), N(15,11)$$

A) $\dfrac{1}{6}$ B) 6

C) $-\dfrac{1}{6}$ D) −6

27) Find the slope of the straight line from the below graph.

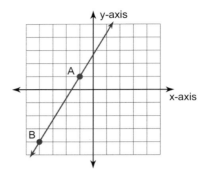

A) $-\dfrac{5}{3}$ B) $-\dfrac{3}{5}$

C) $\dfrac{3}{5}$ D) $\dfrac{5}{3}$

28) Find the slope of the straight line from the below graph.

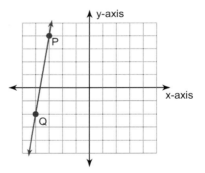

A) 6   B) −6

C) $-\dfrac{1}{6}$ D) $\dfrac{1}{6}$

29) Find the slope of the straight line from the below graph.

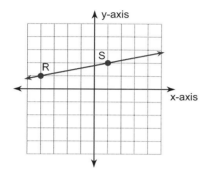

A) 5   B) -5

C) $\dfrac{1}{5}$ D) $-\dfrac{1}{5}$

31) Find the slope of the straight line from the below graph.

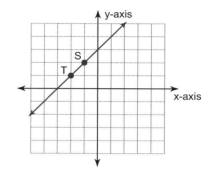

A) 1   B) $-\dfrac{2}{5}$

C) $\dfrac{2}{5}$ D) -1

30) Find the slope of the straight line from the below graph.

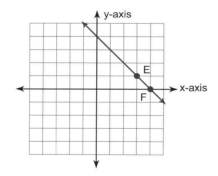

A) -1 B) $-\dfrac{1}{4}$

C) 1   D) $\dfrac{1}{4}$

32) Find the slope of the straight line from the below graph.

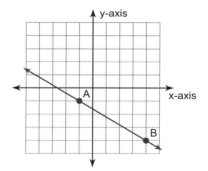

A) $\dfrac{5}{3}$ B) $-\dfrac{5}{3}$

C) $\dfrac{3}{5}$ D) $-\dfrac{3}{5}$

www.math-knots.com | www.a4ace.com

33) Find the slope of the straight line from the below graph.

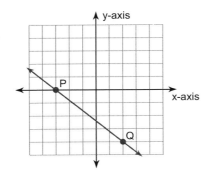

A) $\dfrac{5}{4}$ B) $-\dfrac{5}{4}$

C) $-\dfrac{4}{5}$ D) $\dfrac{4}{5}$

34) Find the slope of the straight line from the below graph.

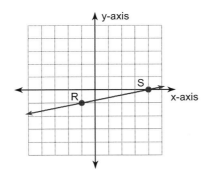

A) $-\dfrac{1}{5}$ B) -5

C) 5 D) $\dfrac{1}{5}$

35) Find the slope of the straight line from the below graph.

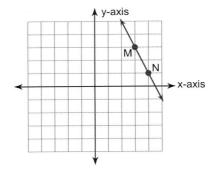

A) 2 B) $-\dfrac{1}{2}$

C) $\dfrac{1}{2}$ D) -2

36) Find the slope of the straight line from the below graph.

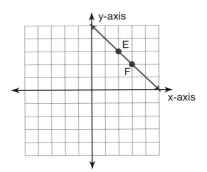

A) 1 B) $-\dfrac{3}{4}$

C) $\dfrac{3}{4}$ D) -1

www.math-knots.com | www.a4ace.com

37) Find the slope of the straight line from the below graph.

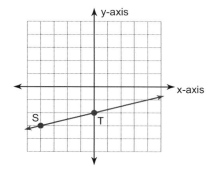

A) $-\dfrac{1}{4}$ B) -4

C) $\dfrac{1}{4}$ D) 4

38) Find the slope of the straight line from the below graph.

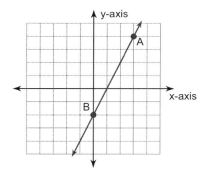

A) -2 B) $\dfrac{1}{2}$

C) $-\dfrac{1}{2}$ D) 2

39) Find the slope of the straight line from the below graph.

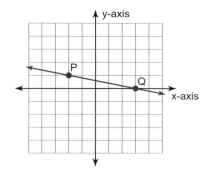

A) 5 B) $\dfrac{1}{5}$

C) -5 D) $-\dfrac{1}{5}$

40) Find the slope of the straight line from the below graph.

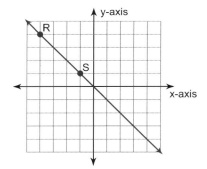

A) 1 B) $-\dfrac{3}{4}$

C) -1 D) $\dfrac{3}{4}$

41) Find the slope of the straight line from the below graph.

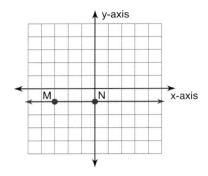

A) 0   B) −1

C) 1   D) Undefined

43) Find the slope of the straight line from the below graph.

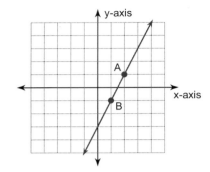

A) $\frac{1}{2}$ B) 2

C) −2   D) $-\frac{1}{2}$

42) Find the slope of the straight line from the below graph.

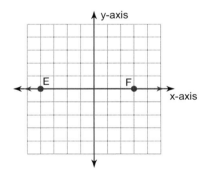

A) Undefined   B) 0

C) 3   D) −3

44) Find the slope of the straight line from the below graph.

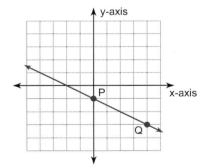

A) −2   B) $-\frac{1}{2}$

C) 2   D) $\frac{1}{2}$

45) Find the slope of the straight line from the below graph.

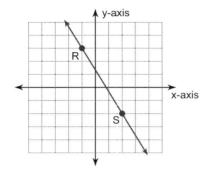

A) $\dfrac{5}{3}$ B) $\dfrac{3}{5}$

C) $-\dfrac{3}{5}$ D) $-\dfrac{5}{3}$

47) Find the slope of the straight line from the below graph.

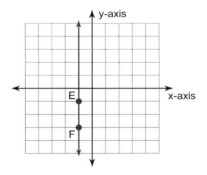

A) $-\dfrac{5}{2}$ B) 0

C) Undefined   D) $\dfrac{5}{2}$

46) Find the slope of the straight line from the below graph.

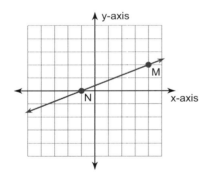

A) $-\dfrac{5}{2}$ B) $\dfrac{2}{5}$

C) $-\dfrac{2}{5}$ D) $\dfrac{5}{2}$

48) Find the slope of the straight line from the below graph.

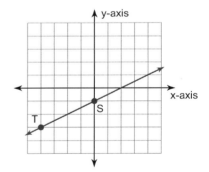

A) -2 B) 2

C) $\dfrac{1}{2}$ D) $-\dfrac{1}{2}$

49) Find the slope of the straight line from the below graph.

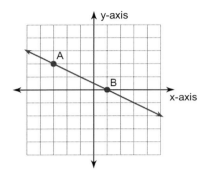

A) $\dfrac{1}{2}$ B) $-\dfrac{1}{2}$

C) 2   D) -2

50) Find the slope of the straight line from the below graph.

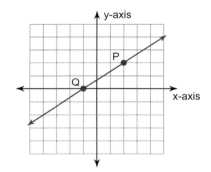

A) $-\dfrac{2}{3}$ B) $-\dfrac{3}{2}$

C) $\dfrac{3}{2}$ D) $\dfrac{2}{3}$

51) Find the slope of the straight line from the below graph.

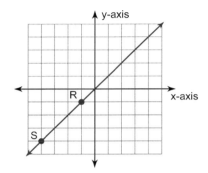

A) $-\dfrac{2}{5}$ B) 1

C) $\dfrac{2}{5}$ D) -1

52) Find the slope of the straight line from the below graph.

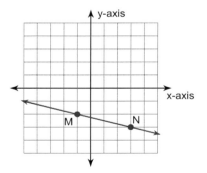

A) $-\dfrac{1}{4}$ B) $\dfrac{1}{4}$

C) 4   D) -4

53) Find the missing coordinate, given the slope

S (49 , – 48) and T (x , – 23); slope: – 1

A) 41   B) 24

C) – 23   D) – 5

54) Find the missing coordinate, given the slope

A (– 7 , – 41) and B (x , 28); slope: $\frac{23}{4}$

A) – 33   B) 25

C) – 16   D) 5

55) Find the missing coordinate, given the slope

R (– 37 , y) and S (– 47 , 30); slope: $-\frac{1}{10}$

A) 40   B) 26

C) – 7   D) 29

56) Find the missing coordinate, given the slope

M (x , – 22) , N (2 , – 34); slope: $\frac{4}{7}$

A) – 28   B) – 26

C) 23   D) 43

57) Find the missing coordinate, given the slope

A (– 42 , – 38) and B (– 2 , y); slope: $\frac{1}{5}$

A) 48   B) – 30

C) 19   D) – 6

58) Find the missing coordinate, given the slope

P (– 44 , – 6) and Q (– 29 , y); slope: $-\frac{14}{5}$

A) 26   B) –48

C) 28   D) –41

59) Find the missing coordinate, given the slope

R (21 , – 27) and S (x , 17); slope: – 4

A) 14   B) –50

C) 42   D) 10

60) Find the missing coordinate, given the slope

M (x , 17) and N (4 , – 23); slope: – 10

A) 46   B) 30

C) 0   D) – 21

61) Find the missing coordinate, given the slope

S (35 , y) and T (42 , – 41) ; slope : 0

A) – 46   B) – 49

C) 30   D) – 41

62) Find the missing coordinate, given the slope

A (x , – 10) and B (– 7 , 6); slope: $\dfrac{16}{3}$

A) – 30   B) – 10

C) 13   D) – 11

63) Find the missing coordinate, given the slope

R (18 , 48) and S (x , – 18) ; slope: $-\dfrac{22}{3}$

A) 16   B) 27

C) – 23   D) 24

64) Find the missing coordinate, given the slope

E (– 6 , 39) and F (x , – 3) ; slope: $\dfrac{6}{5}$

A) – 43   B) – 24

C) – 41   D) 47

65) Find the missing coordinate, given the slope

A (– 25 , 6) and B (x , – 17) ; slope: $-\dfrac{23}{6}$

A) – 44   B) – 4

C) 32   D) – 19

66) Find the missing coordinate, given the slope

P (– 23 , y) and Q (– 37 , – 45) ; slope: 0

A) 22   B) – 45

C) 6   D) 29

67) Find the missing coordinate, given the slope

R (x , 17) and S (20 , 23) ; slope: $-\dfrac{3}{7}$

A) 13   B) 34

C) – 25   D) – 8

68) Find the missing coordinate, given the slope

M (– 26 , 15) and N (– 27 , y) ; slope: 0

A) 35   B) – 49

C) – 27   D) 15

69) Find the missing coordinate, given the slope

S (− 12 , y) and T (− 33 , − 1); slope : $\dfrac{9}{7}$

   A) 4   B) 26

   C) 37   D) − 36

70) Find the missing coordinate, given the slope

A (− 27 , − 18) and B (− 34 , y); slope: $-\dfrac{61}{7}$

   A) − 14   B) 2

   C) 24   D) 43

71) Find the missing coordinate, given the slope

R (x , − 8) and S (5 , 49); slope: undefined

   A) 16   B) 27

   C) − 23   D) 5

72) Find the missing coordinate, given the slope

E (4 , y) and F (36 , − 6) ; slope: $-\dfrac{9}{8}$

   A) 6   B) 18

   C) 30   D) 5

73) Find the missing coordinate, given the slope

A (x , 15) and B (− 45 , − 21); slope: 1

   A) − 29   B) 2

   C) − 9   D) 34

74) Find the missing coordinate, given the slope

P (x , 41) and Q (− 14 , 30); slope: $-\dfrac{11}{8}$

   A) 11   B) − 22

   C) − 19   D) 26

75) Find the missing coordinate, given the slope

R (x , − 45) and S (36 , 25); slope: $\dfrac{7}{8}$

   A) − 44   B) − 35

   C) 26   D) 16

76) Find the missing coordinate, given the slope

S (29 , y) and T (21 , 32); slope : $\dfrac{3}{4}$

   A) − 38   B) − 24

   C) 17   D) 38

Algebra 1

77) Find the missing coordinate, given the slope

A (39 , y) and B (31 , 10) ; slope: $\dfrac{15}{4}$

A) − 12   B) − 50

C) − 39   D) 40

78) Find the missing coordinate, given the slope

R (− 19 , − 44) and S (x , − 14); slope: 15

A) 22   B) − 17

C) 43   D) − 38

79) What is the slope of the straight line given below?

$$y = -3x + 1$$

A) $-\dfrac{1}{3}$ B) $\dfrac{1}{3}$

C) −3   D) 3

80) What is the slope of the straight line given below?

$$y = \dfrac{7}{3}x - 3$$

A) $\dfrac{3}{7}$ B) $-\dfrac{3}{7}$

C) $\dfrac{7}{3}$ D) $-\dfrac{7}{3}$

81) What is the slope of the straight line given below?

$$y = x + 1$$

A) $\dfrac{1}{2}$ B) $-\dfrac{1}{2}$

C) −1   D) 1

82) What is the slope of the straight line given below?

$$y = -2x - 2$$

A) 2   B) $\dfrac{1}{2}$

C) $-\dfrac{1}{2}$ D) −2

83) What is the slope of the straight line given below?

$$y = -\dfrac{1}{4}x - 2$$

A) $\dfrac{1}{4}$ B) 4

C) − 4   D) $-\dfrac{1}{4}$

84) What is the slope of the straight line given below?

$$y = -x$$

A) 1   B) $\dfrac{3}{4}$

C) − 1   D) $-\dfrac{3}{4}$

www.math-knots.com | www.a4ace.com

85) What is the slope of the straight line given below?

$$y = -\frac{1}{2}x$$

A) $-\frac{1}{2}$ B) 2

C) -2 D) $\frac{1}{2}$

86) What is the slope of the straight line given below?

$$y = \frac{5}{4}x - 3$$

A) $-\frac{4}{5}$ B) $\frac{5}{4}$

C) $-\frac{5}{4}$ D) $\frac{4}{5}$

87) What is the slope of the straight line given below?

$$y = 4x + 4$$

A) $\frac{1}{4}$ B) -4

C) $-\frac{1}{4}$ D) 4

88) What is the slope of the straight line given below?

$$y = \frac{3}{2}x + 2$$

A) $\frac{2}{3}$ B) $-\frac{2}{3}$

C) $-\frac{3}{2}$ D) $\frac{3}{2}$

89) What is the slope of the straight line given below?

$$y = 2x + 1$$

A) $-\frac{1}{2}$ B) $\frac{1}{2}$

C) 2   D) -2

90) What is the slope of the straight line given below?

$$y = -\frac{3}{5}x - 2$$

A) $\frac{3}{5}$ B) $-\frac{3}{5}$

C) $\frac{5}{3}$ D) $-\frac{5}{3}$

91) What is the slope of the straight line given below?

$$3x - 5y = 5$$

A) $\frac{3}{5}$ B) $-\frac{5}{3}$

C) $\frac{5}{3}$ D) $-\frac{3}{5}$

92) What is the slope of the straight line given below?

$$5x - 4y = -4$$

A) $-\frac{4}{5}$ B) $\frac{5}{4}$

C) $-\frac{5}{4}$ D) $\frac{4}{5}$

www.math-knots.com | www.a4ace.com

 **Algebra 1**

93) What is the slope of the straight line given below?

$$5x - y = -5$$

A) $-\dfrac{1}{5}$ B) -5

C) 5 D) $\dfrac{1}{5}$

94) What is the slope of the straight line given below?

$$9x + 2y = 8$$

A) $-\dfrac{9}{2}$ B) $\dfrac{2}{9}$

C) $\dfrac{9}{2}$ D) $-\dfrac{2}{9}$

95) What is the slope of the straight line given below?

$$4x + 3y = -12$$

A) $\dfrac{3}{4}$ B) $-\dfrac{3}{4}$

C) $-\dfrac{4}{3}$ D) $\dfrac{4}{3}$

96) What is the slope of the straight line given below?

$$2x - 3y = -5$$

A) $\dfrac{3}{2}$ B) $-\dfrac{2}{3}$

C) $-\dfrac{3}{2}$ D) $\dfrac{2}{3}$

97) What is the slope of the straight line given below?

$$4x + y = -2$$

A) $\dfrac{1}{4}$ B) -4

C) $-\dfrac{1}{4}$ D) 4

98) What is the slope of the straight line given below?

$$8x - 3y = -12$$

A) $\dfrac{3}{8}$ B) $\dfrac{8}{3}$

C) $-\dfrac{3}{8}$ D) $-\dfrac{8}{3}$

99) What is the slope of the straight line given below?

$$3x - y = -1$$

A) $\dfrac{1}{3}$ B) 3

C) -3 D) $-\dfrac{1}{3}$

100) What is the slope of the straight line given below?

$$x = 5$$

A) $\dfrac{4}{3}$ B) $-\dfrac{4}{3}$

C) 0 D) Undefined

www.math-knots.com | www.a4ace.com

101) What is the slope of the straight line given below?

$$6x + 3y = -3$$

102) What is the slope of the straight line given below?

$$y - 1 = -x$$

103) What is the slope of the straight line given below?

$$0 = -4x - 3y$$

104) What is the slope of the straight line given below?

$$-2x = 3 + y$$

105) What is the slope of the straight line given below?

$$0 = x + 10 - 5y$$

106) What is the slope of the straight line given below?

$$-3x = -2 - y$$

107) What is the slope of the straight line given below?

$$0 = 2 - 2x - y$$

108) What is the slope of the straight line given below?

$$\frac{1}{3}y = x + 1$$

109) What is the slope of the straight line given below?

$$10 - x = -5y$$

110) What is the slope of the straight line given below?

$$18x = 15y - 75$$

111) What is the slope of the straight line given below?

$$-x = -2y$$

112) What is the slope of the straight line given below?

$$-y - \frac{4}{5}x = 0$$

113) What is the slope of the straight line given below?

$$-x - 8 = -4y$$

114) What is the slope of the straight line given below?

$$-4y = 6x + 20$$

115) What is the slope of the straight line given below?

$$0 = -x + 2$$

116) What is the slope of the straight line given below?

$$1 - y = -\frac{3}{4}x$$

117) What is the slope of the straight line given below?

$$-y = 2 - 5x$$

118) What is the slope of the straight line given below?

$$\frac{1}{2}x + \frac{1}{3}y = -1$$

119) What is the slope of the straight line given below?

$$y + x = 0$$

120) What is the slope of the straight line given below?

$$8 + 3x = 2y$$

121) Find the y-intercept for the below straight line

$$y = \frac{1}{5}x - 5$$

122) Find the y-intercept for the below straight line

$$y = -\frac{2}{5}x - 2$$

123) Find the y-intercept for the below straight line

$$y = \frac{1}{4}x + 5$$

124) Find the y-intercept for the below straight line

$$y = -\frac{10}{3}x - 5$$

125) Find the x-intercept for the below straight line

$$x - y = 1$$

126) Find the y-intercept for the below straight line

$$y = -2x$$

127) Find the y-intercept for the below straight line

$$y = x - 3$$

128) Find the y-intercept for the below straight line

$$y = \frac{3}{5}x + 2$$

129) Find the y-intercept for the below straight line

$$y = \frac{3}{2}x - 2$$

130) Find the x-intercept for the below straight line

$$2x + 3y = -3$$

131) Find the x-intercept for the below straight line

$$11x - 3y = -21$$

132) Find the x-intercept for the below straight line

$$8x - y = 40$$

133) Find the x-intercept for the below straight line

$$3x - 2y = 4$$

134) Find the y-intercept and x-intercept of the below

$$y = x^2 - 4x - 1$$

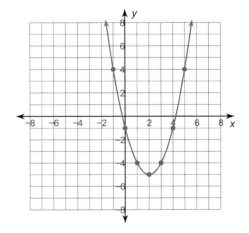

135) Find the x-intercept for the below straight line

$$5x - 6y = -36$$

136) Find the x-intercept for the below straight line

$$8x + 7y = -28$$

137) Find the x-intercept for the below straight line

$$7x + 13y = 10$$

138) Find the y-intercept and x-intercept of the below

$$y = -x^2 + 10x - 20$$

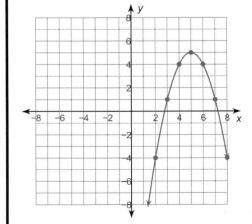

www.math-knots.com | www.a4ace.com

139) Find the y-intercept and
x-intercept of the below

$$y = -|x - 4| + 1$$

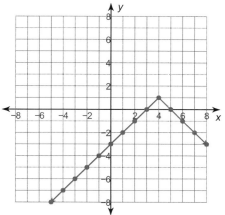

140) Find the y-intercept and
x-intercept of the below

$$y = x^2 + 8x + 13$$

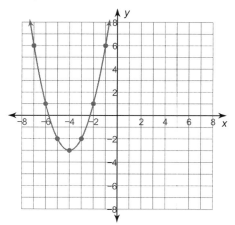

141) Find the y-intercept and
x-intercept of the below

$$y = x^2 - 6x + 4$$

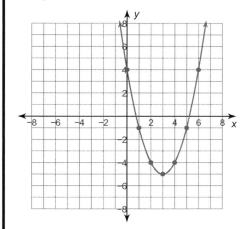

142) Find the y-intercept and
x-intercept of the below

$$y = -3|x - 5| + 7$$

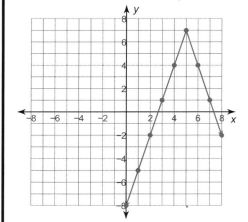

www.math-knots.com | www.a4ace.com

 **Algebra 1**

143) Find the y-intercept and
x-intercept of the below

$$y = -x^2 - 4x - 3$$

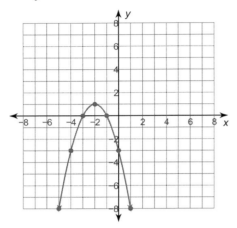

144) Find the y-intercept and
x-intercept of the below

$$2x = -3y$$

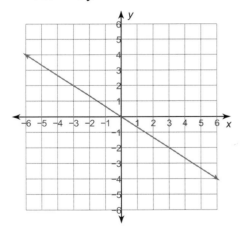

145) Find the y-intercept and
x-intercept of the below

$$y = -x^2 - 8x - 12$$

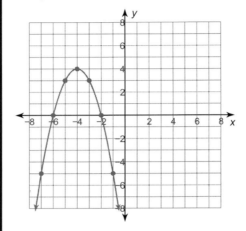

146) Find the y-intercept and
x-intercept of the below

$$-3 = -8x - y$$

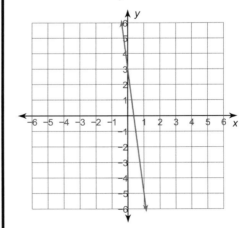

www.math-knots.com | www.a4ace.com

Algebra 1

147) Find the y-intercept and
x-intercept of the below

$$-1 + \frac{1}{3}y - \frac{1}{5}x = 0$$

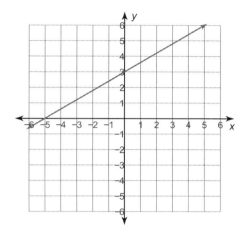

148) Find the y-intercept and
x-intercept of the below

$$y + 2 + \frac{3}{5}x = 0$$

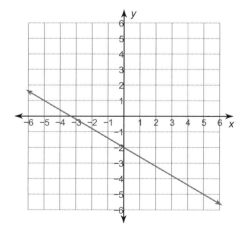

www.math-knots.com | www.a4ace.com

1) Write the equation of the straight line passing through the points A (0 , -1) and B (1 , 4) in slope intercept form.

   A) $y = -5x - 1$ B) $y = 4x - 1$

   C) $y = 5x - 1$ D) $y = -x + 4$

2) Write the equation of the straight line passing through the points A (4 , 0) and B (0 , - 5) in slope intercept form.

   A) $y = \frac{5}{4}x - 5$ B) $y = -\frac{5}{4}x - 5$

   C) $y = \frac{1}{4}x - 5$ D) $y = -\frac{1}{4}x - 5$

3) Write the equation of the straight line passing through the points A (- 3 , - 3) and B (4 , - 4) in slope intercept form.

   A) $y = -\frac{1}{7}x + \frac{24}{7}$ B) $y = \frac{24}{7}x - \frac{1}{7}$

   C) $y = -\frac{1}{7}x - \frac{24}{7}$ D) $y = -\frac{24}{7}x - \frac{1}{7}$

4) Write the equation of the straight line passing through the points A (0 , 5) and B (1 , - 3) in slope intercept form.

   A) $y = -3x + 5$ B) $y = -8x + 5$

   C) $y = -x + 5$ D) $y = 3x + 5$

5) Write the equation of the straight line passing through the points A (3 , 0) and B (0 , - 4) in slope intercept form.

   A) $y = \frac{1}{3}x - 4$ B) $y = \frac{4}{3}x - 4$

   C) $y = -\frac{5}{3}x - 4$ D) $y = -4x - \frac{5}{3}$

6) Write the equation of the straight line passing through the points A (0 , 0) and B (1 , 0) in slope intercept form.

   A) $y = 0$ B) $y = -\frac{2}{3}x$

   C) $y = \frac{2}{3}x$ D) $y = -\frac{1}{3}x$

7) Write the equation of the straight line passing through the points A (- 2 , 0) and B (0 , 2) in slope intercept form.

   A) $y = -2x - 1$ B) $y = -x + 2$

   C) $y = x + 2$ D) $y = 2x - 1$

8) Write the equation of the straight line passing through the points A (- 5 , 5) and B (0 , 5) in slope intercept form.

   A) $x = -1$ B) $y = \frac{1}{3}$

   C) $y = 3x - 1$ D) $y = 5$

9) Write the equation of the straight line passing through the points A (− 3 , 4) and B (− 5 , − 3) in slope intercept form.

A) $y = -\dfrac{29}{2}x - 2$ B) $y = -2x + \dfrac{29}{2}$

C) $y = \dfrac{29}{2}x - 2$ D) $y = \dfrac{7}{2}x + \dfrac{29}{2}$

10) Write the equation of the straight line passing through the points A (− 1 , 5) and B (− 4 , − 3) in slope intercept form.

A) $y = \dfrac{8}{3}x + \dfrac{23}{3}$ B) $y = \dfrac{2}{3}x + \dfrac{23}{3}$

C) $y = -\dfrac{2}{3}x + \dfrac{23}{3}$ D) $y = -x + \dfrac{23}{3}$

11) Write the equation of the straight line passing through the points A (2 , − 4) and B (0 , 1) in slope intercept form.

A) $y = \dfrac{5}{2}x + 1$ B) $y = -2x + 1$

C) $y = 2x + 1$ D) $y = -\dfrac{5}{2}x + 1$

12) Write the equation of the straight line passing through the points A (3 , − 4) and B (0 , 4) in slope intercept form.

A) $y = -\dfrac{8}{3}x + 4$

B) $y = 4x - \dfrac{8}{3}$

C) $y = -\dfrac{5}{3}x - \dfrac{8}{3}$

D) $y = -4x - \dfrac{8}{3}$

13) Write the equation of the straight line passing through the points A (3 , 1) and B (1 , − 3) in slope intercept form.

A) $y = 2x + 5$ B) $y = -5x + 2$

C) $y = 2x - 5$ D) $y = 5x + 2$

14) Write the equation of the straight line passing through the points A (− 4 , 0) and B (0 , − 5) in slope intercept form.

A) $y = -5x - \dfrac{5}{4}$ B) $y = -\dfrac{5}{4}x - 5$

C) $y = 5x - \dfrac{5}{4}$ D) $y = -x - 5$

15) Write the equation of the straight line passing through the points A (− 2 , 0) and B (0 , 1) in slope intercept form.

A) $y = x + \dfrac{1}{2}$ B) $y = -2x + \dfrac{1}{2}$

C) $y = \dfrac{1}{2}x + 1$ D) $y = -\dfrac{1}{2}x + 1$

16) Write the equation of the straight line passing through the points A (3 , − 4) and B (2 , − 2) in slope intercept form.

A) $y = -2x + 2$ B) $y = 2x + 2$

C) $y = -5x - 2$ D) $y = 2x - 2$

17) Write the equation of the straight line passing through the points A (0 , 4) and B (1 , − 1) in slope intercept form.

A) $y = -5x + 4$ B) $y = 4x + 5$

C) $y = -4x + 5$ D) $y = 5x + 4$

18) Write the equation of the straight line passing through the points A (− 3 , − 3) and B (− 3 , 2) in slope intercept form.

A) $y = 3$ B) $x = -3$

C) $y = 3x$ D) $y = -3x$

19) Write the equation of the straight line passing through the points A (3 , 1) and B (4 , − 1) in slope intercept form.

A) $y = -2x + 7$ B) $y = 7x + 3$

C) $y = 3x + 7$ D) $y = 2x + 7$

20) Write the equation of the straight line passing through the points A (− 2 , − 1) and B (− 3 , − 5) in slope intercept form.

A) $y = 4x + 7$ B) $y = 5x + 7$

C) $y = -x + 7$ D) $y = -4x + 7$

21) Write the equation of the straight line passing through the points A (4 , 2) and B (− 3 , − 4) in slope intercept form.

A) $y = -\dfrac{10}{7}x + \dfrac{3}{7}$ B) $y = -\dfrac{3}{7}x - \dfrac{10}{7}$

C) $y = \dfrac{3}{7}x - \dfrac{10}{7}$ D) $y = \dfrac{6}{7}x - \dfrac{10}{7}$

22) Find the slope intercept form of the straight line from the below graph.

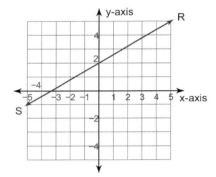

A) $y = \dfrac{3}{5}x + \dfrac{3}{5}$ B) $y = -2x + \dfrac{3}{5}$

C) $y = \dfrac{3}{5}x + 2$ D) $y = 2x + \dfrac{3}{5}$

www.math-knots.com | www.a4ace.com

23) Find the slope intercept form of the straight line from the below graph.

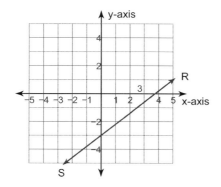

A) $y = \dfrac{1}{5}x + \dfrac{4}{5}$ B) $y = \dfrac{4}{5}x - 3$

C) $y = -3x + \dfrac{4}{5}$ D) $y = \dfrac{3}{5}x + \dfrac{4}{5}$

24) Find the slope intercept form of the straight line from the below graph.

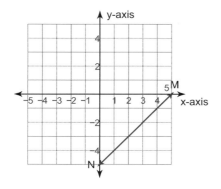

A) $y = 3x - 5$ B) $y = x - 5$

C) $y = -3x - 5$ D) $y = 4x - 5$

25) Find the slope intercept form of the straight line from the below graph.

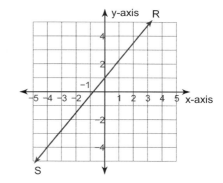

A) $y = x - \dfrac{3}{4}$ B) $y = \dfrac{5}{4}x + 1$

C) $y = -\dfrac{3}{4}x + 1$ D) $y = -x - \dfrac{3}{4}$

26) Find the slope intercept form of the straight line from the below graph.

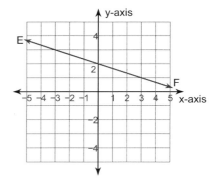

A) $y = 2x - \dfrac{1}{3}$

B) $y = -\dfrac{1}{3}x + 2$

C) $y = -\dfrac{2}{3}x - \dfrac{1}{3}$

D) $y = \dfrac{5}{3}x - \dfrac{1}{3}$

www.math-knots.com | www.a4ace.com

27) Find the slope intercept form of the straight line from the below graph.

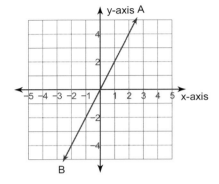

A) $y = 3x$ B) $y = 4x$

C) $y = 2x$ D) $y = -4x$

28) Find the slope intercept form of the straight line from the below graph.

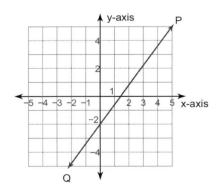

A) $y = x - 2$ B) $y = 3x - 2$

C) $y = \dfrac{7}{5}x - 2$ D) $y = -x - 2$

29) Find the slope intercept form of the straight line from the below graph.

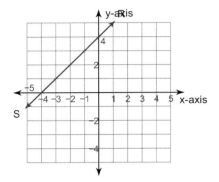

A) $y = -4x - 1$ B) $y = 4x - 1$

C) $y = x + 4$ D) $y = -x + 4$

30) Find the slope intercept form of the straight line from the below graph.

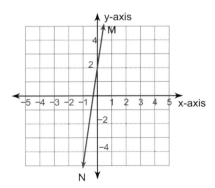

A) $y = 2x + 7$ B) $y = 7x + 2$

C) $y = -7x + 2$ D) $y = 4x + 7$

www.math-knots.com | www.a4ace.com

Algebra 1

31) Find the slope intercept form of the straight line from the below graph.

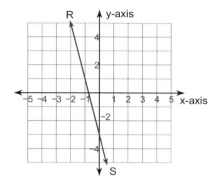

A) $y = x - 3$ B) $y = -3x + 1$

C) $y = -4x - 3$ D) $y = 4x - 3$

32) Find the slope intercept form of the straight line from the below graph.

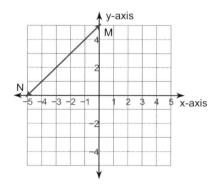

A) $y = -x + 5$ B) $y = x + 5$

C) $y = 3x + 5$ D) $y = -3x + 5$

33) Find the slope intercept form of the straight line from the below graph.

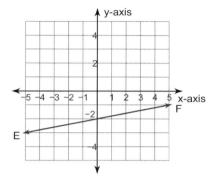

A) $y = x - 2$ B) $y = -2x - 2$

C) $y = -2x + \dfrac{1}{5}$ D) $y = \dfrac{1}{5}x - 2$

34) Find the slope intercept form of the straight line from the below graph.

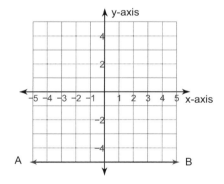

A) $x = -5$ B) $y = -5$

C) $x = 5$ D) $x = 1$

35) Find the slope intercept form of the straight line from the below graph.

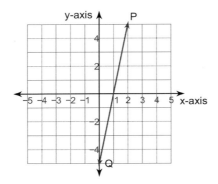

A) $y = 5x - 5$ B) $y = -5x + 5$

C) $y = 5x + 5$ D) $y = 4x + 5$

37) Find the slope intercept form of the straight line from the below graph.

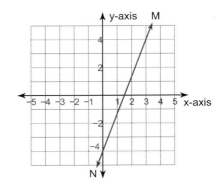

A) $y = \frac{1}{3}x - 4$ B) $y = -\frac{1}{3}x - 4$

C) $y = -4x - \frac{1}{3}$ D) $y = \frac{8}{3}x - 4$

36) Find the slope intercept form of the straight line from the below graph.

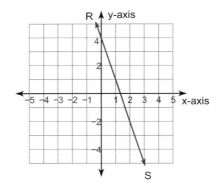

A) $y = 4x - 3$ B) $y = -3x - 4$

C) $y = -3x + 4$ D) $y = -4x - 3$

38) Find the slope intercept form of the straight line from the below graph.

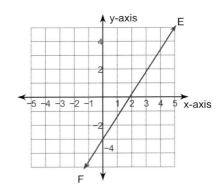

A) $y = -\frac{8}{5}x - 3$ B) $y = -3x - \frac{1}{5}$

C) $y = \frac{8}{5}x - 3$ D) $y = -\frac{1}{5}x - 3$

www.math-knots.com | www.a4ace.com

39) Find the slope intercept form of the straight line from the below graph.

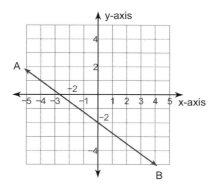

A) $y = -2x - \dfrac{3}{4}$ B) $y = -\dfrac{3}{4}x - \dfrac{5}{4}$

C) $y = -\dfrac{5}{4}x - \dfrac{3}{4}$ D) $y = -\dfrac{3}{4}x - 2$

40) Find the slope intercept form of the straight line from the below graph.

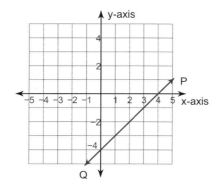

A) $y = x - 4$ B) $y = 4x + 1$

C) $y = -x - 4$ D) $y = -4x + 1$

41) Find the slope intercept form of the straight line from the below graph.

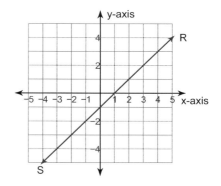

A) $y = -x + 1$ B) $y = x - 1$

C) $y = x - 4$ D) $y = -4x + 1$

42) Find the slope intercept form of the straight line from the below graph.

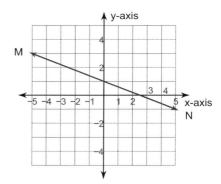

A) $y = -\dfrac{2}{5}x + 1$ B) $y = -\dfrac{4}{5}x + 1$

C) $y = x - 1$ D) $y = -x + 1$

www.math-knots.com | www.a4ace.com

43) Find the slope intercept form of the straight line from the below graph.

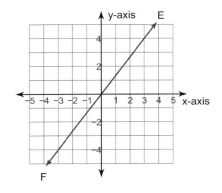

A) $y = -1$

B) $y = -\dfrac{4}{3}x$

C) $y = -x$

D) $y = \dfrac{4}{3}x$

44) Find the slope intercept form of the straight line from the below graph.

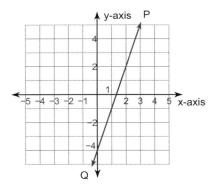

A) $y = x - 4$

B) $y = -3x - 4$

C) $y = -4x + 1$

D) $y = 3x - 4$

45) Find the slope intercept form of the straight line from the below point slope form of the straight line.

through: $(-2, 5)$, slope $= -\dfrac{9}{2}$

A) $y = \dfrac{9}{2}x - 4$

B) $y = -\dfrac{9}{2}x - 4$

C) $y = 2x - 4$

D) $y = -2x - 4$

46) Find the slope intercept form of the straight line from the below point slope form of the straight line.

through: $(-2, -5)$, slope $= \dfrac{1}{2}$

A) $y = -4x - 2$

B) $y = -\dfrac{1}{2}x - 4$

C) $y = \dfrac{1}{2}x - 4$

D) $y = -2x - 4$

47) Find the slope intercept form of the straight line from the below point slope form of the straight line.

through: $(-5, -5)$, slope $= \dfrac{3}{5}$

A) $y = -\dfrac{3}{5}x - 2$

B) $y = \dfrac{3}{5}x - 2$

C) $y = -2x - \dfrac{3}{5}$

D) $y = \dfrac{3}{5}x - \dfrac{3}{5}$

48) Find the slope intercept form of the straight line from the below point slope form of the straight line.

through: $(-1, 2)$, slope $= -5$

A) $y = -3x - 3$

B) $y = -5x - 3$

C) $y = 3x - 3$

D) $y = 5x - 3$

49) Find the slope intercept form of the straight line from the below point slope form of the straight line.

through: $(4 , 3)$, slope = $\dfrac{5}{4}$

A) $y = \dfrac{5}{4}x - 2$ B) $y = -\dfrac{5}{4}x - 2$

C) $y = -x - 2$ D) $y = -2x + \dfrac{5}{4}$

50) Find the slope intercept form of the straight line from the below point slope form of the straight line.

through: $(5 , -2)$, slope = undefined

A) $y = -5$ B) $y = 5x$

C) $y = -5x$ D) $x = 5$

51) Find the slope intercept form of the straight line from the below point slope form of the straight line.

through: $(-3 , 2)$, slope = 1

A) $y = 3x + 5$ B) $y = 4x + 5$

C) $y = -4x + 5$ D) $y = x + 5$

52) Find the slope intercept form of the straight line from the below point slope form of the straight line.

through: $(-3 , 3)$, slope = $\dfrac{2}{5}$

A) $y = \dfrac{2}{5}x + \dfrac{21}{5}$ B) $y = -\dfrac{4}{5}x + \dfrac{21}{5}$

C) $y = \dfrac{21}{5}x - \dfrac{4}{5}$ D) $y = -\dfrac{21}{5}x - \dfrac{4}{5}$

53) Find the slope intercept form of the straight line from the below point slope form of the straight line.

through: $(2, -2)$, slope = -3

A) $y = -3x + 4$ B) $y = -x + 4$

C) $y = x + 4$ D) $y = 3x + 4$

54) Find the slope intercept form of the straight line from the below point slope form of the straight line.

through: $(-3 , -4)$, slope = $\dfrac{2}{3}$

A) $y = x - 2$ B) $y = -x - 2$

C) $y = \dfrac{2}{3}x - 2$ D) $y = -2x + \dfrac{2}{3}$

55) Find the slope intercept form of the straight line from the below point slope form of the straight line.

through: $(-4 , 4)$, slope = $-\dfrac{8}{7}$

A) $y = \dfrac{2}{7}x - \dfrac{4}{7}$ B) $y = -\dfrac{2}{7}x - \dfrac{4}{7}$

C) $y = \dfrac{8}{7}x - \dfrac{4}{7}$ D) $y = -\dfrac{8}{7}x - \dfrac{4}{7}$

56) Find the slope intercept form of the straight line from the below point slope form of the straight line.

through: $(1 , 4)$, slope = 5

A) $y = -x + 5$ B) $y = -x - 5$

C) $y = -5x - 1$ D) $y = 5x - 1$

57) Find the slope intercept form of the straight line from the below point slope form of the straight line.

through: (1 , 1) , slope = 0

A) $y = 1$

B) $y = -\dfrac{4}{5}x + \dfrac{1}{5}$

C) $y = -\dfrac{1}{5}x + \dfrac{1}{5}$

D) $y = \dfrac{1}{5}x - \dfrac{4}{5}$

58) Find the slope intercept form of the straight line from the below point slope form of the straight line.

through: (- 5, 4) , slope = - 1

A) $y = -x - 1$

B) $y = -3x - 1$

C) $y = -x - 3$

D) $y = 3x - 1$

59) Find the slope intercept form of the straight line from the below point slope form of the straight line.

through: (- 4 , 4) , slope = $-\dfrac{7}{4}$

A) $y = -3x - \dfrac{7}{4}$

B) $y = -\dfrac{7}{4}x - 3$

C) $y = -\dfrac{1}{4}x - \dfrac{7}{4}$

D) $y = 3x - \dfrac{7}{4}$

60) Find the slope intercept form of the straight line from the below point slope form of the straight line.

through: (- 5 , - 1) , slope = $\dfrac{1}{5}$

A) $y = -\dfrac{5}{2}x + \dfrac{1}{2}$

B) $y = \dfrac{1}{5}x$

C) $y = -\dfrac{3}{2}x + \dfrac{1}{2}$

D) $y = \dfrac{1}{5}$

61) Find the slope intercept form of the straight line from the below point slope form of the straight line.

through: (- 5, - 5) , slope = $\dfrac{9}{5}$

A) $y = 4x + \dfrac{1}{5}$

B) $y = -4x + \dfrac{1}{5}$

C) $y = \dfrac{1}{5}x + 4$

D) $y = \dfrac{9}{5}x + 4$

62) Find the slope intercept form of the straight line from the below point slope form of the straight line.

through: (- 5 , 2) , slope = $-\dfrac{1}{3}$

A) $y = -\dfrac{1}{3}x - \dfrac{1}{3}$

B) $y = \dfrac{1}{3}x - \dfrac{1}{3}$

C) $y = -x - \dfrac{1}{3}$

D) $y = -\dfrac{1}{3}x + \dfrac{1}{3}$

63) Find the slope intercept form of the straight line from the below point slope form of the straight line.

through: (4 , - 3) , slope = $-\dfrac{7}{4}$

A) $y = x + 4$

B) $y = -\dfrac{5}{4}x + 4$

C) $y = -\dfrac{7}{4}x + 4$

D) $y = \dfrac{7}{4}x + 4$

64) Find the slope intercept form of the straight line from the below point slope form of the straight line.

through: (- 5 , - 5) , slope = 1

A) $y = 1$

B) $y = x$

C) $x = - 1$

D) $x = 1$

www.math-knots.com | www.a4ace.com

65) Find the slope intercept form of the straight line from the below point slope form of the straight line.

through: $(-2, 2)$, slope $= -1$

A) $y = x$ 　　　 B) $y = -1$

C) $x = -1$ 　　　 D) $y = -x$

66) Find the slope intercept form of the straight line from the below point slope form of the straight line.

through: $(5, -3)$, slope $= -1$

A) $y = -3x + 2$ 　　 B) $y = 3x + 2$

C) $y = -x + 2$ 　　 D) $y = 2x + 3$

67) Find the slope intercept form of the straight line from the below point standard form of the straight line.

$$y - 4 = -(x + 1)$$

A) $y = -x + 3$ 　　 B) $y = -3x - 1$

C) $y = 3x - 1$ 　　 D) $y = -x - 3$

68) Find the slope intercept form of the straight line from the below point standard form of the straight line.

$$y + 1 = \frac{3}{2}(x + 2)$$

A) $y = -\frac{3}{2}x + 2$ 　　 B) $y = -2x + \frac{3}{2}$

C) $y = 2x + \frac{3}{2}$ 　　 D) $y = \frac{3}{2}x + 2$

69) Find the slope intercept form of the straight line from the below point standard form of the straight line.

$$y - 2 = -\frac{1}{4}(x - 4)$$

A) $y = -\frac{1}{4}x + 3$ 　　 B) $y = x - \frac{1}{4}$

C) $y = -x - \frac{1}{4}$ 　　 D) $y = 3x - \frac{1}{4}$

70) Find the slope intercept form of the straight line from the below point standard form of the straight line.

$$y + 4 = -5(x - 1)$$

A) $y = x - 4$ 　　 B) $y = -4x + 1$

C) $y = -5x + 1$ 　　 D) $y = 5x + 1$

71) Find the slope intercept form of the straight line from the below point standard form of the straight line .

$$y - 2 = -\frac{5}{9}(x + 5)$$

A) $y = \frac{5}{9}x - \frac{7}{9}$ 　　 B) $y = \frac{4}{9}x - \frac{7}{9}$

C) $y = -\frac{5}{9}x - \frac{7}{9}$ 　 D) $y = -\frac{1}{3}x - \frac{7}{9}$

72) Find the slope intercept form of the straight line from the below point standard form of the straight line .

$$y + 4 = \frac{6}{5}(x + 3)$$

A) $y = \frac{6}{5}x - \frac{2}{5}$ 　　 B) $y = \frac{2}{5}x - \frac{6}{5}$

C) $y = -\frac{2}{5}x - \frac{6}{5}$ 　 D) $y = -\frac{6}{5}x - \frac{2}{5}$

 　　 　　 www.math-knots.com | www.a4ace.com

73) Find the slope intercept form of the straight line from the below point standard form of the straight line.

$$y - 3 = 0$$

A) $y = 3$

B) $y = -x + \dfrac{1}{4}$

C) $y = \dfrac{1}{4}x - \dfrac{3}{4}$

D) $y = -\dfrac{3}{4}x + \dfrac{1}{4}$

74) Find the slope intercept form of the straight line from the below point standard form of the straight line.

$$y + 3 = 0$$

A) $y = -\dfrac{1}{3}$

B) $x = -1$

C) $x = 1$

D) $y = -3$

75) Find the slope intercept form of the straight line from the below point standard form of the straight line.

$$y - 1 = -\dfrac{3}{4}(x + 4)$$

A) $y = -\dfrac{3}{4}x - 2$

B) $y = -2x - \dfrac{1}{4}$

C) $y = -\dfrac{1}{4}x - 2$

D) $y = \dfrac{1}{2}x - \dfrac{1}{4}$

76) Find the slope intercept form of the straight line from the below point standard form of the straight line.

$$y - 3 = -\dfrac{1}{5}x$$

A) $y = -\dfrac{1}{5}x - \dfrac{4}{5}$

B) $y = -\dfrac{1}{5}x + 3$

C) $y = -\dfrac{4}{5}x - \dfrac{1}{5}$

D) $y = 3x - \dfrac{1}{5}$

77) Find the slope intercept form of the straight line from the below point standard form of the straight line .

$$y - 5 = -\dfrac{3}{5}(x + 3)$$

A) $y = -\dfrac{16}{5}x - \dfrac{3}{5}$

B) $y = \dfrac{16}{5}x - \dfrac{3}{5}$

C) $y = x - \dfrac{3}{5}$

D) $y = -\dfrac{3}{5}x + \dfrac{16}{5}$

78) Find the slope intercept form of the straight line from the below point standard form of the straight line.

$$y = 0$$

A) $x = -1$

B) $y = \dfrac{1}{3}$

C) $y = 0$

D) $y = 1$

328   www.math-knots.com | www.a4ace.com

79) Find the slope intercept form of the straight line from the below point standard form of the straight line.

$$y - 2 = -\frac{7}{4}(x + 4)$$

A) $y = \frac{7}{4}x - 5$ B) $y = -\frac{7}{4}x - 5$

C) $y = -\frac{3}{4}x - 5$ D) $y = -5x + \frac{7}{4}$

80) Find the slope intercept form of the straight line from the below point standard form of the straight line.

$$y - 1 = \frac{2}{5}(x - 3)$$

A) $y = x - \frac{1}{5}$ B) $y = \frac{1}{5}x - \frac{1}{5}$

C) $y = -\frac{1}{5}x - \frac{1}{5}$ D) $y = \frac{2}{5}x - \frac{1}{5}$

81) Find the slope intercept form of the straight line from the below point standard form of the straight line.

$$y - 1 = -\frac{4}{3}(x + 3)$$

A) $y = -\frac{4}{3}x - \frac{4}{3}$ B) $y = \frac{4}{3}x - \frac{4}{3}$

C) $y = -\frac{4}{3}x - 3$ D) $y = -3x - \frac{4}{3}$

82) Find the slope intercept form of the straight line from the below point standard form of the straight line.

$$y - 3 = \frac{5}{3}(x - 3)$$

A) $y = -2x + \frac{5}{3}$ B) $y = \frac{5}{3}x - 2$

C) $y = \frac{5}{3}x - \frac{4}{3}$ D) $y = -\frac{4}{3}x + \frac{5}{3}$

83) Find the slope intercept form of the straight line from the below point standard form of the straight line .

$$y + 1 = 2(x - 2)$$

A) $y = -5x - 5$ B) $y = 4x - 5$

C) $y = 2x - 5$ D) $y = -2x - 5$

84) Find the slope intercept form of the straight line from the below point standard form of the straight line.

$$y - 4 = -2(x + 2)$$

A) $y = -2x$ B) $y = 4x + 2$

C) $y = 2$ D) $y = 2x$

85) Find the slope intercept form of the straight line from the below point standard form of the straight line.

$$y + 1 = 2 (x - 1)$$

A) $y = -2x - 3$ B) $y = 4x - 3$

C) $y = 2x - 3$ D) $y = -4x - 3$

86) Find the slope intercept form of the straight line from the below point standard form of the straight line.

$$y = - (x + 1)$$

A) $y = -x + 3$ B) $y = -x - 1$

C) $y = 3x - 1$ D) $y = 5x + 3$

87) Find the slope intercept form of the straight line from the below point standard form of the straight line.

$$0 = x + 4$$

A) $x = -4$ B) $y = 4$

C) $y = 4x$ D) $y = -x$

88) Find the slope intercept form of the straight line from the below point standard form of the straight line.

$$y + 5 = \frac{2}{3}(x + 3)$$

A) $y = \frac{5}{3}x - 3$ B) $y = -\frac{2}{3}x - 3$

C) $y = -3x + \frac{5}{3}$ D) $y = \frac{2}{3}x - 3$

89) Find the slope intercept form of the straight line from the below point standard form of the straight line .

$$3x - y = -7$$

A) $y = -7x + 3$ B) $y = 3x + 7$

C) $y = 3x - 7$ D) $y = 7x + 3$

90) Find the slope intercept form of the straight line from the below point standard form of the straight line.

$$16x + y = 8$$

A) $y = -8x + 16$ B) $y = -16x + 8$

C) $y = 8x + 16$ D) $y = 16x + 8$

91) Find the slope intercept form of the straight line from the below point standard form of the straight line.

$$4x + y = -2$$

A) $y = -3x - 2$ B) $y = -4x - 2$

C) $y = -2x - 3$ D) $y = x - 3$

94) Find the slope intercept form of the straight line from the below point standard form of the straight line.

$$y = -3$$

A) $y = 3x$ B) $y = -3x$

C) $y = -3$ D) $y = 5x$

92) Find the slope intercept form of the straight line from the below point standard form of the straight line.

$$3x + 4y = -4$$

A) $y = \frac{3}{4}x - 1$ B) $y = -\frac{3}{4}x - 1$

C) $y = -x + \frac{3}{4}$ D) $y = x + \frac{3}{4}$

95) Find the slope intercept form of the straight line from the below point standard form of the straight line .

$$x + y = 0$$

A) $y = x$ B) $y = -1$

C) $y = -x$ D) $y = 1$

93) Find the slope intercept form of the straight line from the below point standard form of the straight line.

$$8x - 5y = -25$$

A) $y = -\frac{1}{5}x + 5$ B) $y = -\frac{8}{5}x + 5$

C) $y = 5x + \frac{8}{5}$ D) $y = \frac{8}{5}x + 5$

96) Find the slope intercept form of the straight line from the below point standard form of the straight line.

$$2x + 7y = -33$$

A) $y = \frac{2}{7}x - \frac{33}{7}$ B) $y = \frac{4}{7}x - \frac{33}{7}$

C) $y = -\frac{2}{7}x - \frac{33}{7}$ D) $y = -\frac{4}{7}x - \frac{33}{7}$

97) Find the slope intercept form of the straight line from the below point standard form of the straight line.

$$9x - 4y = 51$$

A) $y = \dfrac{51}{4}x - \dfrac{9}{4}$ B) $y = -\dfrac{51}{4}x - \dfrac{9}{4}$

C) $y = -\dfrac{9}{4}x - \dfrac{51}{4}$ D) $y = \dfrac{9}{4}x - \dfrac{51}{4}$

98) Find the slope intercept form of the straight line from the below point standard form of the straight line.

$$x + 7y = -8$$

A) $y = \dfrac{1}{7}x - \dfrac{8}{7}$ B) $y = -\dfrac{3}{7}x - \dfrac{8}{7}$

C) $y = -\dfrac{8}{7}x - \dfrac{1}{7}$ D) $y = -\dfrac{1}{7}x - \dfrac{8}{7}$

99) Find the slope intercept form of the straight line from the below point standard form of the straight line.

$$14x - 5y = 30$$

A) $y = \dfrac{2}{5}x + \dfrac{14}{5}$ B) $y = \dfrac{14}{5}x + \dfrac{2}{5}$

C) $y = \dfrac{14}{5}x - 6$ D) $y = -6x + \dfrac{14}{5}$

100) Find the slope intercept form of the straight line from the below point standard form of the straight line.

$$y = 6$$

A) $y = \dfrac{1}{6}x$ B) $x = -1$

C) $y = 6$ D) $y = -\dfrac{2}{3}x$

101) Find the slope intercept form of the straight line from the below point standard form of the straight line .

$$2x + y = 7$$

A) $y = 2x + 7$ B) $y = 7x - 2$

C) $y = -2x + 7$ D) $y = -4x - 2$

102) Find the slope intercept form of the straight line from the below point standard form of the straight line.

$$9x + 8y = 24$$

A) $y = \dfrac{5}{4}x + 3$ B) $y = \dfrac{1}{4}x + 3$

C) $y = -\dfrac{1}{4}x + 3$ D) $y = -\dfrac{9}{8}x + 3$

103) Find the slope intercept form of the straight line from the below point standard form of the straight line.

$$5x - 8y = -4$$

A) $y = -\dfrac{5}{8}x + \dfrac{1}{2}$ B) $y = \dfrac{1}{2}x - \dfrac{5}{8}$

C) $y = -\dfrac{1}{2}x - \dfrac{5}{8}$ D) $y = \dfrac{5}{8}x + \dfrac{1}{2}$

104) Find the slope intercept form of the straight line from the below point standard form of the straight line.

$$11x - 8y = -24$$

A) $y = 3x - 1$ B) $y = \dfrac{11}{8}x + 3$

C) $y = \dfrac{1}{4}x + 3$ D) $y = -x + 3$

105) Find the slope intercept form of the straight line from the below point standard form of the straight line.

$$12x - 7y = 42$$

A) $y = -\dfrac{12}{7}x - 6$ B) $y = \dfrac{12}{7}x - 6$

C) $y = \dfrac{2}{7}x - \dfrac{12}{7}$ D) $y = -6x - \dfrac{12}{7}$

106) Find the slope intercept form of the straight line from the below point standard form of the straight line.

$$3x + y = -7$$

A) $y = 3x - 7$ B) $y = -x - 7$

C) $y = -3x - 7$ D) $y = x - 7$

107) Write the equation of the straight line passing through the points A (3 , -4) and B (-3 , -3) in point slope form.

A) $y + 4 = -\dfrac{1}{6}(x - 3)$

B) $y - 3 = -21(x - 4)$

C) $y - 4 = -6(x - 3)$

D) $y + 3 = 6(x + 4)$

108) Write the equation of the straight line passing through the points A (− 3 , − 2) and B (0 , 2) in point slope form.

A) $0 = x + 2$

B) $y - 2 = -\dfrac{4}{3}(x - 3)$

C) $y - 3 = -\dfrac{4}{3}(x - 2)$

D) $y + 2 = \dfrac{4}{3}(x + 3)$

109) Write the equation of the straight line passing through the points A (− 2 , 4) and B (− 5 , 1) in point slope form.

A) $y + 2 = -(x + 4)$

B) $y + 4 = x - 2$

C) $y - 4 = x + 2$

D) $y + 4 = -(x + 2)$

110) Write the equation of the straight line passing through the points A (− 1 , 4) and B (− 5 , 5) in point slope form.

A) $y - 4 = x - 1$

B) $y - 4 = -\dfrac{1}{4}(x + 1)$

C) $0 = x - 4$

D) $0 = x - 1$

111) Write the equation of the straight line passing through the points A (0 , 5) and B (− 1 , − 1) in point slope form.

A) $y = 4 (x - 5)$ B) $y - 5 = x$

C) $y = -4 (x - 5)$ D) $y - 5 = 6 x$

112) Write the equation of the straight line passing through the points A (− 2 , − 2) and B (0 , 5) in point slope form.

A) $y + 2 = 5(x + 2)$

B) $y + 2 = -\dfrac{5}{2}(x + 2)$

C) $y + 2 = -5(x + 2)$

D) $y + 2 = \dfrac{7}{2}(x + 2)$

113) Write the equation of the straight line passing through the points A (4 , − 2) and B (0 , 5) in point slope form.

A) $y - 4 = \dfrac{4}{5}(x + 2)$

B) $y + 4 = \dfrac{2}{25}(x - 2)$

C) $y + 2 = -\dfrac{7}{4}(x - 4)$

D) $y + 4 = -\dfrac{4}{25}(x + 2)$

114) Write the equation of the straight line passing through the points A (− 5 , 5) and B (− 1 , 1) in point slope form.

A) $y - 5 = - (x + 5)$

B) $y + 5 = 2 (x + 5)$

C) $y - 5 = 2 (x + 5)$

D) $y - 5 = x + 5$

115) Write the equation of the straight line passing through the points A (3 , 0) and B (4 , − 2) in point slope form.

A) $y = -2(x - 3)$

B) $y - 3 = -2x$

C) $y = -2(x + 3)$

D) $y + 3 = 3x$

116) Write the equation of the straight line passing through the points A (2 , 1) and B (− 5 , − 4) in point slope form.

A) $y - 1 = \frac{5}{7}(x + 2)$

B) $y + 2 = \frac{4}{7}(x - 1)$

C) $y - 1 = \frac{5}{7}(x - 2)$

D) $y - 1 = \frac{3}{7}(x - 2)$

117) Write the equation of the straight line passing through the points A (0 , − 4) and B (5, 1) in point slope form.

A) $y + 4 = x$ B) $y = 4(x - 4)$

C) $y = -(x + 4)$ D) $y = x + 4$

118) Write the equation of the straight line passing through the points A (0 , 2) and B (− 3 , 4) in point slope form.

A) $y = \frac{3}{5}(x - 2)$

B) $y - 2 = -\frac{2}{3}x$

C) $y = -\frac{3}{5}(x + 2)$

D) $y = \frac{3}{5}(x + 2)$

119) Write the equation of the straight line passing through the points A (5 , 4) and B (3 , 2) in point slope form.

A) $y - 4 = \frac{1}{4}(x - 5)$

B) $y - 4 = -\frac{11}{4}(x - 5)$

C) $y + 4 = -\frac{11}{4}(x - 5)$

D) $y - 5 = -\frac{1}{3}(x - 4)$

120) Write the equation of the straight line passing through the points A (0 , 3) and B (− 1 , − 1) in point slope form.

A) $y + 3 = -2x$

B) $y - 3 = 3x$

C) $y = -3(x + 3)$

D) $y - 3 = \frac{3}{5}x$

121) Write the equation of the straight line passing through the points A (2 , − 4) and B (1 , − 4) in point slope form.

   A) $y + 4 = - (x + 2)$

   B) $y + 4 = x + 2$

   C) $y - 2 = x - 4$

   D) $y + 4 = 0$

122) Write the equation of the straight line passing through the points A (0 , 5) and B (− 1 , − 5) in point slope form.

   A) $y + 5 = x$

   B) $y + 5 = 10 x$

   C) $y = - (x + 5)$

   D) $y - 5 = 10 x$

123) Write the equation of the straight line passing through the points A (4 , − 3) and B (5 , 0) in point slope form.

   A) $y - 4 = 3 (x - 3)$

   B) $y + 3 = 3 (x - 4)$

   C) $y + 4 = \frac{1}{5} (x + 3)$

   D) $y + 4 = -\frac{1}{5} (x + 3)$

124) Write the equation of the straight line passing through the points A (− 2 , 2) and B (− 3 , 1) in point slope form.

   A) $y - 2 = x + 2$

   B) $y - 2 = -\frac{5}{4} (x + 2)$

   C) $y + 2 = \frac{5}{4} (x - 2)$

   D) $y - 2 = \frac{5}{4} (x + 2)$

125) Write the equation of the straight line passing through the points A (− 4 , 5) and B (− 5 , 2) in point slope form.

   A) $y + 5 = 3 (x + 4)$

   B) $y - 5 = -\frac{1}{3} (x + 4)$

   C) $y - 5 = 3 (x + 4)$

   D) $y + 4 = -\frac{1}{3} (x - 5)$

126) Write the equation of the straight line passing through the points A (5 , 0) and B (0 , − 5) in point slope form.

   A) $y = x - 5$

   B) $y = - 4 (x - 5)$

   C) $y = - (x - 5)$

   D) $y = - (x + 5)$

www.math-knots.com | www.a4ace.com

127) Write the equation of the straight line passing through the points A (4 , − 3) and B (4, 4) in point slope form.

A) $0 = x − 4$

B) $y − 3 = 0$

C) $y − 3 = 3 (x + 4)$

D) $y + 3 = 0$

128) Write the equation of the straight line passing through the points A (− 2 , 0) and B (− 2 , − 3) in point slope form.

A) $y = 0$

B) $0 = x + 2$

C) $0 = x$

D) $y − 2 = 0$

129) Write the equation of the straight line passing through the point A (2 , − 2) with a slope of slope = − 3 in point slope form.

A) $y + 2 = − 3 (x − 2)$

B) $y − 2 = \dfrac{1}{3}(x − 2)$

C) $y − 2 = −\dfrac{1}{3}(x − 2)$

D) $y − 2 = 3 (x − 2)$

130) Write the equation of the straight line passing through the point A (1 , 1) with a slope of slope = 2 in point slope form.

A) $y + 1 = − 4 (x + 1)$

B) $y − 1 = x − 1$

C) $y − 1 = − (x − 1)$

D) $y − 1 = 2 (x − 1)$

131) Write the equation of the straight line passing through the point A (5 , − 4) with a slope of slope $= −\dfrac{2}{9}$ in point slope form.

A) $y − 4 = \dfrac{5}{26}(x + 5)$

B) $y + 5 = −\dfrac{5}{2}(x − 4)$

C) $y + 4 = −\dfrac{2}{9}(x − 5)$

D) $y − 4 = \dfrac{9}{2}(x − 5)$

132) Write the equation of the straight line passing through the point A (− 4 , − 2) with a slope of slope = 1 in point slope form.

A) $y + 4 = x + 2$

B) $y + 2 = − (x − 4)$

C) $y + 2 = x + 4$

D) $y − 2 = − (x − 4)$

133) Write the equation of the straight line passing through the point A (3 , 2) with a slope of slope = $\frac{5}{3}$ in point slope form.

A) $y + 3 = \frac{3}{5}(x + 2)$

B) $y + 3 = \frac{5}{3}(x + 2)$

C) $y - 2 = \frac{5}{3}(x - 3)$

D) $y + 3 = \frac{2}{5}(x + 2)$

134) Write the equation of the straight line passing through the point A (− 1 , 2) with a slope of slope = $-\frac{7}{6}$ in point slope form.

A) $y + 2 = \frac{6}{7}(x + 1)$

B) $y + 2 = -\frac{6}{7}(x + 1)$

C) $y - 2 = -\frac{7}{6}(x + 1)$

D) $y + 1 = -\frac{7}{6}(x + 2)$

135) Write the equation of the straight line passing through the point A (4 , 4) with a slope of slope = $\frac{5}{4}$ in point slope form.

A) $y - 4 = -\frac{4}{5}(x - 4)$

B) $y + 4 = -\frac{4}{5}(x - 4)$

C) $y - 4 = \frac{4}{5}(x + 4)$

D) $y - 4 = \frac{5}{4}(x - 4)$

136) Write the equation of the straight line passing through the point A (1 , 2) with a slope of slope = − 3 in point slope form.

A) $y - 1 = 3 (x - 2)$

B) $y - 2 = \frac{1}{2}(x - 1)$

C) $y - 2 = - 3 (x - 1)$

D) $y - 1 = 3 (x + 2)$

137) Write the equation of the straight line passing through the point A (− 3 , − 5) with a slope of slope = 7 in point slope form.

A) $y - 3 = - 16 (x - 5)$

B) $y + 3 = x - 5$

C) $y + 3 = 7 (x - 5)$

D) $y + 5 = 7 (x + 3)$

138) Write the equation of the straight line passing through the point P(0 , 5) with a slope of 9 in point slope form.

A) $y - 3 = - 9 (x - 5)$

B) $y - 5 = 9x$

C) $y + 3 = 0 (x - 9)$

D) $y - 9 = 5 (x - 0)$

139) Write the equation of the straight line passing through the point $A(-4, -3)$ with a slope of slope $= \frac{5}{4}$ in point slope form.

A) $y + 3 = \frac{5}{4}(x + 4)$

B) $y + 4 = \frac{5}{4}(x + 3)$

C) $y - 4 = 4(x + 3)$

D) $y - 4 = -\frac{1}{4}(x + 3)$

140) Write the equation of the straight line passing through the point $A(3, 5)$ with a slope of slope $= \frac{9}{5}$ in point slope form.

A) $y + 3 = -\frac{9}{5}(x - 5)$

B) $y + 5 = \frac{9}{5}(x - 3)$

C) $y - 5 = \frac{9}{5}(x - 3)$

D) $y - 3 = -\frac{9}{5}(x - 5)$

141) Write the equation of the straight line passing through the point $A(5, 2)$ with a slope of slope $= \frac{1}{5}$ in point slope form.

A) $y - 5 = -5(x - 2)$

B) $y - 5 = -5(x + 2)$

C) $y - 2 = \frac{1}{5}(x - 5)$

D) $y + 5 = 5(x - 2)$

142) Write the equation of the straight line passing through the point $A(5, -4)$ with a slope of slope $= \frac{1}{5}$ in point slope form.

A) $y + 5 = 4(x + 4)$

B) $y + 4 = \frac{1}{5}(x - 5)$

C) $y - 5 = 5(x + 4)$

D) $y + 4 = -\frac{1}{5}(x - 5)$

143) Write the equation of the straight line passing through the point $A(0, -4)$ with a slope of slope $= -1$ in point slope form.

A) $y = -\frac{1}{5}(x - 4)$

B) $y = \frac{1}{5}(x - 4)$

C) $y + 4 = -x$

D) $y - 4 = -\frac{1}{5}x$

144) Write the equation of the straight line passing through the point $A(2, 5)$ with a slope of slope $= \frac{7}{2}$ in point slope form.

A) $y - 5 = \frac{2}{7}(x - 2)$

B) $y - 5 = -\frac{2}{7}(x - 2)$

C) $y - 5 = \frac{7}{2}(x - 2)$

D) $y - 2 = \frac{2}{7}(x + 5)$

145) Write the equation of the straight line passing through the point A (1 , − 2) with a slope of slope = $\frac{5}{4}$ in point slope form.

A) y − 2 = 0

B) y − 2 = − (x − 1)

C) y − 1 = 0

D) y + 2 = $\frac{5}{4}$ (x − 1)

146) Write the equation of the straight line passing through the point A (2 ,− 3) with a slope of slope = − $\frac{3}{2}$ in point slope form.

A) y − 2 = − 4 (x − 3)

B) y − 3 = − (x − 2)

C) y + 3 = − $\frac{3}{2}$ (x − 2)

D) y − 3 = $\frac{3}{2}$ (x − 2)

147) Write the equation of the straight line passing through the point A (− 4 , 1) with a slope of slope = $\frac{1}{2}$ in point slope form.

A) y − 4 = − $\frac{1}{2}$ (x + 1)

B) y − 1 = $\frac{1}{2}$ (x + 4)

C) y + 1 = − 2 (x + 4)

D) y − 4 = − 2 (x − 1)

148) Write the equation of the straight line passing through the point A (1 , 1) with a slope of slope = − 1 in point slope form.

A) y − 1 = − (x + 1)

B) y − 1 = − 4 (x + 1)

C) y + 1 = − (x − 1)

D) y − 1 = − (x − 1)

149) Write the equation of the straight line passing through the point A (5 , − 4) with a slope of slope = − $\frac{7}{5}$ in point slope form.

A) y − 5 = $\frac{15}{2}$ (x + 4)

B) y + 5 = $\frac{2}{15}$ (x + 4)

C) y + 4 = − $\frac{7}{5}$ (x − 5)

D) y − 4 = $\frac{2}{15}$ (x − 5)

150) Write the equation of the straight line passing through the point A (4 , − 5) with a slope of slope = − 2 in point slope form.

A) y − 5 = 3 (x − 4)

B) y + 5 = − 2 (x − 4)

C) y − 4 = − 2 (x + 5)

D) y − 5 = − $\frac{1}{3}$ (x − 4)

151) Write the equation of the straight line passing through the point $A(-4, 0)$ with a slope of slope $= -\dfrac{1}{2}$ in point slope form.

   A) $y = -2(x+4)$

   B) $y = 2(x+4)$

   C) $y + 4 = -2x$

   D) $y = -\dfrac{1}{2}(x+4)$

152) Write the equation of the straight line passing through the point $A(-1, -4)$ with a slope of slope $= -1$ in point slope form.

   A) $y - 1 = -\dfrac{1}{5}(x-4)$

   B) $y + 4 = -(x+1)$

   C) $y + 1 = -(x+4)$

   D) $y - 1 = \dfrac{1}{5}(x+4)$

153) Rewrite the below equation of straight line into slope intercept form of the straight line.

$$y + 5 = -\dfrac{7}{3}(x-3)$$

   A) $y = -\dfrac{7}{3}x + 2$ B) $y = \dfrac{2}{3}x + 2$

   C) $y = -\dfrac{2}{3}x + 2$ D) $y = -\dfrac{1}{3}x + 2$

154) Rewrite the below equation of straight line into slope intercept form of the straight line.

$$y + 5 = -\dfrac{4}{5}(x-5)$$

   A) $y = -\dfrac{4}{5}x - 1$ B) $y = \dfrac{2}{5}x - 1$

   C) $y = \dfrac{4}{5}x - 1$ D) $y = \dfrac{3}{5}x - 1$

155) Rewrite the below equation of straight line into slope intercept form of the straight line.

$$y - 4 = -\dfrac{6}{5}(x+5)$$

   A) $y = -2x - \dfrac{4}{5}$

   B) $y = -\dfrac{4}{5}x - 2$

   C) $y = -\dfrac{6}{5}x - 2$

   D) $y = \dfrac{3}{5}x - \dfrac{4}{5}$

156) Rewrite the below equation of straight line into slope intercept form of the straight line.

$$y + 3 = \dfrac{3}{4}(x+4)$$

   A) $y = \dfrac{5}{4}x$ B) $y = -\dfrac{3}{4}$

   C) $y = \dfrac{3}{4}x$ D) $y = -\dfrac{3}{4}x$

157) Rewrite the below equation of straight line into slope intercept form of the straight line.

$$y - 5 = \frac{5}{8}(x - 4)$$

A) $y = \frac{5}{8}x + \frac{5}{2}$ B) $y = -\frac{3}{8}x + \frac{5}{2}$

C) $y = -\frac{5}{8}x + \frac{5}{2}$ D) $y = \frac{5}{2}x - \frac{3}{8}$

158) Rewrite the below equation of straight line into slope intercept form of the straight line.

$$y + 5 = \frac{5}{7}(x + 5)$$

A) $y = \frac{5}{7}x - \frac{10}{7}$ B) $y = \frac{3}{7}x - \frac{10}{7}$

C) $y = -\frac{5}{7}x - \frac{10}{7}$ D) $y = -\frac{1}{7}x - \frac{10}{7}$

159) Rewrite the below equation of straight line into slope intercept form of the straight line.

$$y + 4 = -\frac{5}{4}(x - 4)$$

A) $y = -\frac{5}{4}x + 1$ B) $y = \frac{1}{2}x + 1$

C) $y = x + \frac{1}{2}$ D) $y = \frac{5}{4}x + 1$

160) Rewrite the below equation of straight line into slope intercept form of the straight line.

$$y + 3 = \frac{1}{2}(x + 5)$$

A) $y = -2x + 1$ B) $y = -\frac{1}{2}x + 1$

C) $y = \frac{1}{2}x - \frac{1}{2}$ D) $y = x - \frac{1}{2}$

161) Rewrite the below equation of straight line into slope intercept form of the straight line.

$$y - 2 = -\frac{1}{3}(x + 1)$$

A) $y = \frac{5}{3}x - \frac{1}{3}$ B) $y = -\frac{1}{3}x + \frac{5}{3}$

C) $y = \frac{1}{3}x + \frac{5}{3}$ D) $y = -\frac{5}{3}x - \frac{1}{3}$

162) Rewrite the below equation of straight line into slope intercept form of the straight line.

$$y - 4 = -\frac{3}{2}(x + 2)$$

A) $y = 2x + 1$ B) $y = -2x + 1$

C) $y = -\frac{3}{2}x + 1$ D) $y = \frac{3}{2}x + 1$

163) Rewrite the below equation of straight line into slope intercept form of the straight line.

$$y + 5 = \frac{5}{4}(x - 1)$$

A) $y = -\frac{25}{4}x + \frac{5}{4}$ B) $y = \frac{5}{4}x - \frac{25}{4}$

C) $y = \frac{1}{2}x + \frac{5}{4}$ D) $y = \frac{3}{4}x + \frac{5}{4}$

164) Rewrite the below equation of straight line into slope intercept form of the straight line.

$$y + 5 = -\frac{4}{3}(x + 1)$$

A) $y = x - \frac{19}{3}$ B) $y = -\frac{19}{3}x + 1$

C) $y = -\frac{4}{3}x - \frac{19}{3}$ D) $y = -x - \frac{19}{3}$

165) Rewrite the below equation of straight line into slope intercept form of the straight line.

$$y + 1 = -2(x + 2)$$

A) $y = -5x - 2$ B) $y = 5x - 2$

C) $y = 2x - 2$ D) $y = -2x - 5$

166) Rewrite the below equation of straight line into slope intercept form of the straight line.

$$y + 1 = 0$$

A) $y = 5x - 1$ B) $y = x - 1$

C) $y = -1$ D) $y = -x - 1$

167) Rewrite the below equation of straight line into slope intercept form of the straight line.

$$y - 2 = \frac{1}{4}(x + 4)$$

A) $y = \frac{1}{4}x + 3$ B) $y = -\frac{5}{2}x + 3$

C) $y = \frac{1}{2}x + 3$ D) $y = -\frac{1}{2}x + 3$

168) Rewrite the below equation of straight line into slope intercept form of the straight line.

$$y + 3 = -\frac{5}{4}(x - 4)$$

A) $y = \frac{3}{4}x - \frac{5}{4}$ B) $y = -2x - \frac{5}{4}$

C) $y = 2x - \frac{5}{4}$ D) $y = -\frac{5}{4}x + 2$

169) Rewrite the below equation of straight line into slope intercept form of the straight line.

$$y - 4 = \frac{7}{5}(x - 5)$$

A) $y = \frac{7}{5}x - 3$ B) $y = -\frac{1}{5}x - 3$

C) $y = -\frac{4}{5}x - 3$ D) $y = -\frac{7}{5}x - 3$

170) Rewrite the below equation of straight line into slope intercept form of the straight line.

$$y - 3 = -(x + 3)$$

A) $x = 1$ B) $y = -1$

C) $y = -x$ D) $x = -1$

171) Rewrite the below equation of straight line into slope intercept form of the straight line.

$$y + 1 = x - 2$$

A) $y = 3x - 3$ B) $y = -3x - 3$

C) $y = -3x + 1$ D) $y = x - 3$

172) Rewrite the below equation of straight line into slope intercept form of the straight line.

$$y + 2 = -\frac{1}{8}(x - 5)$$

A) $y = -\frac{11}{8}x - \frac{5}{8}$ B) $y = -\frac{11}{8}x - \frac{1}{8}$

C) $y = -\frac{1}{8}x - \frac{11}{8}$ D) $y = -\frac{5}{8}x - \frac{11}{8}$

173) Rewrite the below equation of straight line into slope intercept form of the straight line.

$$y + 1 = -2(x + 1)$$

A) $y = 2x - 3$ B) $y = 3x + 2$

C) $y = -2x - 3$ D) $y = -3x + 2$

174) Rewrite the below equation of straight line into slope intercept form of the straight line.

$$x + 8y = -8$$

A) $y = \frac{1}{8}x - 1$ B) $y = -x + \frac{1}{8}$

C) $y = -\frac{1}{8}x - 1$ D) $y = \frac{5}{8}x + \frac{1}{8}$

175) Rewrite the below equation of straight line into slope intercept form of the straight line.

$$3x + 4y = -8$$

A) $y = -\dfrac{5}{4}x - 2$ B) $y = -2x - \dfrac{3}{4}$

C) $y = -2x - \dfrac{5}{4}$ D) $y = -\dfrac{3}{4}x - 2$

176) Rewrite the below equation of straight line into slope intercept form of the straight line.

$$x + 2y = 0$$

A) $y = \dfrac{3}{2}x$ B) $y = -\dfrac{1}{2}x$

C) $y = \dfrac{5}{2}$ D) $y = \dfrac{5}{2}x$

177) Rewrite the below equation of straight line into slope intercept form of the straight line.

$$5x - y = -8$$

A) $y = 5x + 8$ B) $y = x + 5$

C) $y = -x + 5$ D) $y = 8x + 5$

178) Rewrite the below equation of straight line into slope intercept form of the straight line.

$$8x - 7y = 0$$

A) $y = \dfrac{8}{7}x$ B) $x = -7$

C) $y = \dfrac{7}{5}$ D) $y = \dfrac{8}{7}$

179) Rewrite the below equation of straight line into slope intercept form of the straight line.

$$x - y = 0$$

A) $y = x$ B) $y = -x + 5$

C) $y = 5x$ D) $y = 5$

180) Rewrite the below equation of straight line into slope intercept form of the straight line.

$$4x - 3y = -8$$

A) $y = \dfrac{5}{3}x + \dfrac{8}{3}$ B) $y = \dfrac{8}{3}x - \dfrac{5}{3}$

C) $y = -\dfrac{5}{3}x + \dfrac{8}{3}$ D) $y = \dfrac{4}{3}x + \dfrac{8}{3}$

181) Rewrite the below equation of straight line into slope intercept form of the straight line.

$$8x + 9y = 32$$

A) $y = -\dfrac{1}{3}x + \dfrac{32}{9}$ B) $y = -\dfrac{8}{9}x + \dfrac{32}{9}$

C) $y = \dfrac{2}{9}x + \dfrac{32}{9}$ D) $y = \dfrac{8}{9}x + \dfrac{32}{9}$

182) Rewrite the below equation of straight line into slope intercept form of the straight line.

$$x = -8$$

A) $x = -8$ B) $y = -8x$

C) $y = 8x$ D) $y = 8$

183) Rewrite the below equation of straight line into slope intercept form of the straight line.

$$11x - 2y = 10$$

A) $y = \dfrac{1}{2}x - 5$ B) $y = \dfrac{11}{2}x - 5$

C) $y = -5x + \dfrac{11}{2}$ D) $y = -\dfrac{11}{2}x - 5$

184) Rewrite the below equation of straight line into slope intercept form of the straight line.

$$2x + y = -1$$

A) $y = 2x - 1$ B) $y = -2x - 1$

C) $y = -x + 2$ D) $y = -3x - 1$

185) Rewrite the below equation of straight line into slope intercept form of the straight line.

$$15x + y = -67$$

A) $y = 67x + 4$ B) $y = 4x - 67$

C) $y = -67x + 4$ D) $y = -15x - 67$

186) Rewrite the below equation of straight line into slope intercept form of the straight line.

$$10x + 7y = -49$$

A) $y = -\dfrac{4}{7}x - \dfrac{10}{7}$ B) $y = -\dfrac{5}{7}x - \dfrac{10}{7}$

C) $y = -\dfrac{10}{7}x - 7$ D) $y = -7x - \dfrac{10}{7}$

187) Rewrite the below equation of straight line into slope intercept form of the straight line.

$$y = -4$$

A) $y = \frac{1}{4}x - 1$ B) $y = \frac{5}{4}x - 1$

C) $y = -\frac{1}{4}x - 1$ D) $y = -4$

188) Rewrite the below equation of straight line into slope intercept form of the straight line.

$$x = 5$$

A) $x = -5$ B) $y = -1$

C) $y = -5$ D) $x = 5$

189) Rewrite the below equation of straight line into slope intercept form of the straight line.

$$7x - 5y = 0$$

A) $y = 4x$ B) $y = x$

C) $y = \frac{7}{5}x$ D) $y = -4x$

190) Rewrite the below equation of straight line into slope intercept form of the straight line.

$$4x + 3y = 6$$

A) $y = -\frac{4}{3}x + 2$

B) $y = \frac{4}{3}x + 2$

C) $y = 2x + \frac{4}{3}$

D) $y = -2x + \frac{4}{3}$

191) Rewrite the below equation of straight line into slope intercept form of the straight line.

$$x - y = 5$$

A) $y = -5x + 3$ B) $y = 3x - 5$

C) $y = -3x - 5$ D) $y = x - 5$

192) Rewrite the below equation of straight line into slope intercept form of the straight line.

$$x + 3y = -9$$

A) $y = \frac{1}{3}x - \frac{1}{3}$ B) $y = \frac{1}{3}x - 3$

C) $y = -3x - \frac{1}{3}$ D) $y = -\frac{1}{3}x - 3$

193) Rewrite the below equation of straight line into slope intercept form of the straight line.

$$x = -2$$

A) $x = -2$ B) $y = 5x - 2$

C) $y = -5x - 2$ D) $y = \dfrac{2}{5}$

194) Rewrite the below equation of straight line into slope intercept form of the straight line.

$$12x + 7y = -28$$

A) $y = -\dfrac{12}{7}x - 4$ B) $y = 4x + \dfrac{1}{7}$

C) $y = \dfrac{1}{7}x - 4$ D) $y = -4x + \dfrac{1}{7}$

195) Rewrite the below equation of straight line into slope intercept form of the straight line.

$$4x + 3y = 12$$

A) $y = -\dfrac{1}{3}x - \dfrac{4}{3}$ B) $y = -\dfrac{4}{3}x + 4$

C) $y = -\dfrac{4}{3}x - \dfrac{1}{3}$ D) $y = 4x - \dfrac{4}{3}$

196) Rewrite the below equation of straight line into slope intercept form of the straight line.

$$x - 5y = -40$$

A) $y = \dfrac{1}{5}x + 8$

B) $y = 8x - \dfrac{3}{5}$

C) $y = -\dfrac{4}{5}x + 8$

D) $y = -\dfrac{3}{5}x + 8$

197) Rewrite the below equation of straight line into slope intercept form of the straight line.

$$x + 3y = -18$$

A) $y = -\dfrac{1}{3}x - 6$ B) $y = \dfrac{1}{3}x - 6$

C) $y = 6x + \dfrac{1}{3}$ D) $y = -6x + \dfrac{1}{3}$

198) Rewrite the below equation of straight line into slope intercept form of the straight line.

$$7x - y = 8$$

A) $y = 7x - 8$ B) $y = -3x - 8$

C) $y = -7x - 8$ D) $y = -x - 8$

199) Write the equation of the straight line passing through the points A (-3 , -5) and B (-2 , 3) in standard form.

   A) $3x - y = 2$ B) $x + 3y = 2$

   C) $8x - y = -19$ D) $5x + 3y = -12$

200) Write the equation of the straight line passing through the points A (-5 , 3) and B (4 , 0) in standard form.

   A) $3x - y = -4$ B) $x + 3y = 4$

   C) $3x + y = 4$ D) $3x + y = -4$

201) Write the equation of the straight line passing through the points A (-5 , -2) and B (4 , -1) in standard form.

   A) $x + 9y = 13$ B) $3x + 9y = -13$

   C) $x - 9y = 13$ D) $x + 9y = -13$

202) Write the equation of the straight line passing through the points A (3 , 0) and B (4 , -2) in standard form.

   A) $2x + 5y = 2$ B) $2x + y = 6$

   C) $5x + 2y = -2$ D) $x - 2y = 6$

203) Write the equation of the straight line passing through the points A (-2 , -5) and B (-4 , -1) in standard form.

   A) $2x + y = -9$ B) $2x - y = -9$

   C) $5x + y = -2$ D) $x + 4y = -2$

204) Write the equation of the straight line passing through the points A (4 , -3) and B (0 , 2) in standard form.

   A) $5x + 4y = 8$ B) $x + y = 2$

   C) $x + y = -2$ D) $x - y = 2$

205) Write the equation of the straight line passing through the points A (0 , 4) and B (-2 , -5) in standard form.

   A) $9x - 2y = -8$ B) $2x + 9y = -8$

   C) $4x + 9y = 8$ D) $5x + 9y = 8$

206) Write the equation of the straight line passing through the points A (0 , -3) and B (-4 , -2) in standard form.

   A) $x - y = 3$ B) $x + 4y = -12$

   C) $3x + y = -1$ D) $x + y = -3$

Algebra 1

207) Write the equation of the straight line passing through the points A (1 , 5) and B (5 , 3) in standard form.

A) $2x - 4y = 11$ B) $x - 2y = 11$

C) $5x - 4y = -11$ D) $x + 2y = 11$

208) Write the equation of the straight line passing through the points A (4 , – 1) and B (– 5 , 4) in standard form.

A) $5x + 9y = 11$ B) $11x + 9y = 18$

C) $11x - 9y = -2$ D) $18x + 9y = -11$

209) Write the equation of the straight line passing through the points A (4 , – 1) and B (0 , – 3) in standard form.

A) $2x + y = 4$ B) $2x - y = -6$

C) $x - 2y = 6$ D) $2x + y = -6$

210) Write the equation of the straight line passing through the points A (4 , – 1) and B (0 , 5) in standard form.

A) $x + y = 4$ B) $x - y = 5$

C) $3x + 2y = 10$ D) $3x - 2y = 10$

211) Write the equation of the straight line passing through the points A (0 , 0) and B (2 , – 4) in standard form.

A) $2x + y = 5$ B) $x - 2y = 4$

C) $2x - y = 0$ D) $2x + y = 0$

212) Write the equation of the straight line passing through the points A (5 , 5) and B (0 , – 1) in standard form.

A) $5x + 6y = -5$ B) $6x - 5y = 5$

C) $5x + 6y = 5$ D) $2x + 6y = -5$

213) Write the equation of the straight line passing through the points A (– 3 , 4) and B (2 , – 1) in standard form.

A) $4x + y = -8$ B) $x + y = 1$

C) $2x - y = -2$ D) $x - y = 1$

214) Write the equation of the straight line passing through the points A (1 , – 1) and B (0 , 1) in standard form.

A) $x - 2y = -2$ B) $2x + y = -1$

C) $2x - y = -2$ D) $2x + y = 1$

215) Write the equation of the straight line passing through the points A (2 , 4) and B (5 , − 4) in standard form.

    A) $8x + 3y = 28$ B) $9x − 4y = 3$

    C) $4x + 3y = − 9$ D) $6x + 3y = − 8$

216) Write the equation of the straight line passing through the points A (4 , 1) and B (− 3 , − 3) in standard form.

    A) $3x − 7y = − 7$ B) $4x − 7y = 9$

    C) $4x + 7y = 21$ D) $21x + 7y = 4$

217) Write the equation of the straight line passing through the points A (− 2 , − 2) and B (1 , 5) in standard form.

    A) $3x − 7y = − 8$ B) $7x − 3y = − 8$

    C) $5x − 7y = 14$ D) $3x − 7y = 14$

218) Write the equation of the straight line passing through the points A (0 , − 4) and B (− 4 , 4) in standard form.

    A) $2x − y = − 4$ B) $x + 3y = − 1$

    C) $2x + y = − 4$ D) $3x + y = 4$

219) Write the equation of the straight line passing through the points A (4 , 2) and B (0 , 5) in standard form.

    A) $4x = 3$ B) $3x + 4y = 20$

    C) $3x − 4y = − 20$ D) $4x − 3y = 0$

220) Write the equation of the straight line passing through the points A (0 ,− 1) and B (4 , − 3) in standard form.

    A) $2x − y = 1$ B) $x + 2y = − 2$

    C) $x − 4y = − 4$ D) $x + y = − 1$

www.math-knots.com | www.a4ace.com

352   www.math-knots.com | www.a4ace.com

Algebra 1
Answer Keys

354   www.math-knots.com | www.a4ace.com

Algebra 1

Week 1		Week 1		Week 1		Week 1	
1.	B	30.	B	59.	B	88.	D
2.	B	31.	C	60.	A	89.	A
3.	A	32.	D	61.	A	90.	B
4.	A	33.	B	62.	C	91.	D
5.	B	34.	A	63.	C	92.	D
6.	D	35.	B	64.	D	93.	B
7.	D	36.	B	65.	A	94.	A
8.	C	37.	D	66.	D	95.	D
9.	C	38.	D	67.	A	96.	C
10.	C	39.	A	68.	B	97.	A
11.	C	40.	D	69.	B	98.	A
12.	A	41.	B	70.	A	99.	B
13.	A	42.	D	71.	D	100.	C
14.	D	43.	B	72.	C	101.	D
15.	C	44.	B	73.	C	102.	D
16.	B	45.	B	74.	B	103.	D
17.	C	46.	C	75.	A		
18.	B	47.	A	76.	D		
19.	A	48.	B	77.	C		
20.	C	49.	A	78.	D		
21.	B	50.	A	79.	D		
22.	D	51.	B	80.	A		
23.	B	52.	A	81.	A		
24.	A	53.	A	82.	C		
25.	C	54.	D	83.	B		
26.	A	55.	C	84.	A		
27.	A	56.	C	85.	C		
28.	B	57.	B	86.	D		
29.	D	58.	D	87.	A		

 **Algebra 1**

Week 2

1.	7x
2.	34 -7n
3.	20n
4.	-7b
5.	10v - 11
6.	22r - 8
7.	10a
8.	10 - 24n
9.	-27a
10.	22 + 25v
11.	-2v
12.	3k
13.	36 + 14x
14.	-30 + 20n
15.	15x
16.	-2 + 21v
17.	30n
18.	-4a
19.	14n
20.	29x - 10
21.	9x
22.	-3x
23.	-4x
24.	-30n
25.	r + 6
26.	-12 - 6a
27.	6.6n - 3.6
28.	m - 6.881
29.	-1.2m

Week 2

30.	0.1r + 4.2
31.	-8.8 + 0.1p
32.	2.2n - 8.61
33.	v - 10.95
34.	-9.6x
35.	2.8x
36.	1+ 16.7m
37.	11k
38.	6.5p
39.	12.389 - 12x
40.	-19.6m
41.	-2.39n
42.	-5.9x
43.	-7.4x
44.	-18.5k
45.	-10.4n
46.	3.4 - 9.7k
47.	-5.6n
48.	3n + 7.2
49.	8.7x
50.	-3.9p
51.	p - 6.4
52.	-4x
53.	9n - 63
54.	7r - 7
55.	48n - 60
56.	2n - 16
57.	-40r + 40
58.	2b - 18

Week 2

59.	10 + 30k
60.	3- 27x
61.	-9 - 72n
62.	-1 - 6x
63.	72n - 7n
64.	6k - 6
65.	30 + 20x
66.	10n + 100
67.	-5 + 35x
68.	5 - 5m
69.	-28x - 4
70.	-2 - 8x
71.	-90k - 27
72.	-30 - 9x
73.	18n + 63
74.	32k - 64
75.	-10r + 100
76.	-9m + 6
77.	90m -19
78.	-3x - 9
79.	10x + 60
80.	-3.33x -24.642
81.	6.1 - 9.15n
82.	-4.79n + 11.017
83.	-75.525b + 48.45
84.	0.7n - 2.52
85.	-6.825x - 9.555
86.	-5.28n + 13.28
87.	7.6 - 11.4b

                   www.math-knots.com | www.a4ace.com

 **Algebra 1**

Week 3

1.	-11.56 + 40.8v
2.	-11.22 + 0.68k
3.	0.4 - 1.68m
4.	2.3 + 17.71x
5.	-18.62 + 9.386m
6.	-1.6v -12.16
7.	68.985a - 42.33
8.	-9.6x - 11.52
9.	5.3n - 12.826
10.	-2.3v - 8.51
11.	-6.1v + 9.76
12.	-0.6x - 3.636
13.	25.35 - 32.37b
14.	-41.4n - 34.04
15.	-42.885x + 46.697
16.	-10n + 3
17.	-35.88m - 4.68
18.	48.84 - 72.52x
19.	34.2a - 81.7
20.	-44.16 - 85.44n
21.	-31 + 36a
22.	7n + 24
23.	72r - 84
24.	63n - 110
25.	8 + 55x
26.	4n + 27
27.	-23x - 5
28.	32n + 60
29.	64n + 88

Week 3

30.	-67 + 77n
31.	-42
32.	18 - 72b
33.	-2x - 33
34.	-37 - 88x
35.	31k + 20
36.	-11n - 33
37.	-15n + 12
38.	-12m - 108
39.	-13a + 12
40.	54 - 2b
41.	28r + 62
42.	-44r + 120
43.	2x - 25
44.	121n - 77
45.	4k + 8
46.	71r + 54
47.	18 + 47x
48.	7r + 23
49.	-42 - 120x
50.	-27b - 152
51.	-45r - 74
52.	-63n + 98
53.	21m - 9
54.	59p + 103
55.	-38a + 52
56.	-154 - 34k
57.	72x - 189
58.	34n + 49

Week 3

59.	-6r + 59
60.	-132r + 102
61.	-78p - 102
62.	-52x - 4
63.	74n + 10
64.	-93n + 83
65.	-8p + 2
66.	13x + 84
67.	92 - 8n
68.	136n - 208
69.	-9k - 74
70.	-107n - 27
71.	14a - 54
72.	-120n + 26
73.	-54 + 58x
74.	-92a - 78
75.	$-\frac{41}{20} + \frac{9}{16}v$
76.	$-\frac{6}{5}r + \frac{89}{30}$
77.	$-\frac{233}{48} - \frac{55}{16}b$
78.	$\frac{4}{3}m - \frac{7}{8}$
79.	$2p + \frac{3}{25}$
80.	$-\frac{9}{2}n + 7$
81.	$\frac{1}{8} - \frac{15}{8}n$
82.	10v

www.math-knots.com | www.a4ace.com

Algebra 1

Week 3	
83.	$-\dfrac{1}{12}v + \dfrac{3}{4}$
84.	$-\dfrac{127}{20}x - \dfrac{13}{2}$
85.	$-\dfrac{63}{8} + \dfrac{17}{4}r$
86.	$-\dfrac{5}{n} - \dfrac{97}{12}$
87.	$-\dfrac{5}{2}b - \dfrac{7}{2}$
88.	$-\dfrac{203}{36} + \dfrac{5}{3}x$
89.	$-\dfrac{21}{4} + \dfrac{5}{4}v$
90.	$-\dfrac{29}{20}n + \dfrac{15}{8}$
91.	$\dfrac{14}{5} - 6x$
92.	$\dfrac{1}{2}x + \dfrac{47}{40}$
93.	$-\dfrac{25}{8}n - \dfrac{9}{8}$
94.	$-3x + 1$
95.	$-\dfrac{7}{3}b - \dfrac{91}{12}$
96.	$-\dfrac{34}{5} - \dfrac{19}{10}x$
97.	$-\dfrac{91}{60}b + \dfrac{5}{4}$
98.	$\dfrac{121}{30} - \dfrac{3}{4}x$
99.	$-\dfrac{13}{4} - \dfrac{3}{4}$
100.	$12 - \dfrac{81}{20}p$
101.	$-\dfrac{19}{6} + \dfrac{4}{5}r$
102.	$-\dfrac{8}{15}a - 1$

Week 4	
1.	D
2.	C
3.	D
4.	C
5.	D
6.	A
7.	A
8.	C
9.	D
10.	A
11.	A
12.	A
13.	A
14.	D
15.	C
16.	A
17.	C
18.	D
19.	C
20.	C
21.	B
22.	D
23.	A
24.	A
25.	C
26.	C
27.	A
28.	B
29.	D

Week 4	
30.	A
31.	D
32.	B
33.	B
34.	A
35.	B
36.	C
37.	A
38.	C
39.	C
40.	D
41.	C
42.	D
43.	D
44.	C
45.	B
46.	C
47.	D
48.	A
49.	B
50.	A
51.	C
52.	C
53.	C
54.	C
55.	C
56.	B
57.	A
58.	B

Week 4	
59.	D
60.	C
61.	B
62.	D
63.	-3
64.	-8
65.	-7
66.	8
67.	6
68.	-2
69.	3
70.	8
71.	5
72.	-7
73.	-6
74.	4
75.	7
76.	-8
77.	8
78.	-7
79.	-6
80.	5
81.	-6
82.	6
83.	-7
84.	4
85.	-8
86.	8
87.	-7

www.math-knots.com | www.a4ace.com

Algebra 1

Week 4		Week 5		Week 5		Week 5		Week 6	
88.	7	1.	A	30.	D	59.	B	1.	D
89.	-7	2.	C	31.	D	60.	B	2.	D
		3.	A	32.	B	61.	C	3.	C
		4.	C	33.	A	62.	D	4.	A
		5.	A	34.	C	63.	A	5.	B
		6.	B	35.	B	64.	D	6.	D
		7.	B	36.	C	65.	D	7.	A
		8.	D	37.	B	66.	C	8.	D
		9.	B	38.	A	67.	D	9.	D
		10.	D	39.	C	68.	D	10.	D
		11.	B	40.	B	69.	C	11.	D
		12.	C	41.	A	70.	B	12.	C
		13.	D	42.	A	71.	B	13.	B
		14.	C	43.	A	72.	B	14.	C
		15.	B	44.	A	73.	B	15.	C
		16.	A	45.	A	74.	C	16.	A
		17.	B	46.	C	75.	A	17.	C
		18.	B	47.	D	76.	C	18.	C
		19.	C	48.	B	77.	C	19.	A
		20.	C	49.	D	78.	C	20.	D
		21.	A	50.	B	79.	C	21.	C
		22.	C	51.	D	80.	B	22.	B
		23.	B	52.	D	81.	D	23.	A
		24.	D	53.	B	82.	A	24.	C
		25.	A	54.	A	83.	D	25.	B
		26.	A	55.	B	84.	C	26.	D
		27.	B	56.	B	85.	D	27.	D
		28.	D	57.	A			28.	C
		29.	A	58.	B			29.	A

Week 6		Week 6		Week 6		Week 7		Week 7	
30.	D	59.	C	88.	B	1.	D	30.	C
31.	D	60.	B	89.	D	2.	D	31.	C
32.	D	61.	D	90.	B	3.	D	32.	A
33.	B	62.	D	91.	B	4.	B	33.	B
34.	C	63.	A	92.	A	5.	A	34.	A
35.	B	64.	C	93.	B	6.	C	35.	C
36.	D	65.	D	94.	A	7.	C	36.	D
37.	C	66.	A	95.	C	8.	D	37.	A
38.	D	67.	D	96.	C	9.	D	38.	A
39.	B	68.	D	97.	A	10.	D	39.	C
40.	D	69.	C	98.	B	11.	D	40.	D
41.	D	70.	B	99.	B	12.	A	41.	D
42.	D	71.	B	100.	C	13.	B	42.	C
43.	B	72.	D	101.	D	14.	C	43.	A
44.	D	73.	B	102.	D	15.	B	44.	A
45.	D	74.	B	103.	C	16.	D	45.	B
46.	D	75.	C	104.	D	17.	A	46.	C
47.	D	76.	C	105.	B	18.	A	47.	C
48.	C	77.	A	106.	B	19.	B	48.	D
49.	C	78.	D	107.	C	20.	D	49.	C
50.	D	79.	A	108.	C	21.	B	50.	B
51.	D	80.	D	109.	C	22.	D	51.	B
52.	C	81.	D	110.	A	23.	B	52.	A
53.	B	82.	D			24.	B	53.	B
54.	C	83.	C			25.	D	54.	B
55.	D	84.	D			26.	A	55.	A
56.	A	85.	D			27.	C	56.	A
57.	C	86.	D			28.	C	57.	B
58.	D	87.	C			29.	A	58.	A

www.math-knots.com | www.a4ace.com

 **Algebra 1**

Week 7		Week 8		Week 8		Week 8		Week 8	
59.	A	1.	C	30.	B	59.	C	88.	3
60.	D	2.	A	31.	B	60.	B	89.	4.3
61.	A	3.	C	32.	A	61.	B	90.	4.08
62.	A	4.	C	33.	B	62.	D	91.	26.4
63.	D	5.	D	34.	B	63.	B	92.	31.96
64.	D	6.	A	35.	B	64.	B	93.	19.6
65.	D	7.	C	36.	C	65.	B	94.	51
66.	B	8.	B	37.	C	66.	C	95.	31.4
67.	C	9.	A	38.	C	67.	D	96.	17.97
68.	B	10.	A	39.	B	68.	B	97.	17.13
69.	D	11.	C	40.	D	69.	B	98.	17
70.	C	12.	C	41.	D	70.	B	99.	16.8
71.	A	13.	A	42.	B	71.	C	100.	16
72.	D	14.	A	43.	D	72.	B	101.	18
73.	D	15.	D	44.	C	73.	B	102.	2
74.	B	16.	B	45.	B	74.	D	103.	2.57
75.	C	17.	B	46.	A	75.	C	104.	2.44
76.	A	18.	A	47.	A	76.	A	105.	2
77.	B	19.	D	48.	D	77.	6	106.	2.6
78.	A	20.	C	49.	A	78.	5.67	107.	2
79.	A	21.	C	50.	B	79.	4	108.	3
80.	D	22.	D	51.	B	80.	7	109.	1
81.	A	23.	A	52.	C	81.	3	110.	1.43
82.	C	24.	D	53.	B	82.	1.32	111.	1.36
83.	A	25.	C	54.	C	83.	1.25	112.	29.5
84.	A	26.	C	55.	A	84.	44	113.	31.28
85.	B	27.	C	56.	C	85.	44.6	114.	27.2
86.	C	28.	C	57.	B	86.	43	115.	38.9
		29.	C	58.	B	87.	46	116.	11.7

www.math-knots.com | www.a4ace.com

Week 8		Week 8		Week 9		Week 9	
117.	5.66	146.	5.03	1.	A	30.	A
118.	5.4	147.	33.2	2.	D	31.	B
119.	46	148.	33.18	3.	D	32.	-4 - 16r
120.	45.73	149.	29	4.	C	33.	-7x + 46
121.	43	150.	37.5	5.	2 - 4x	34.	B
122.	51	151.	8.5	6.	D	35.	1.8 - 5.7x
123.	8	152.	6.15	7.	B	36.	C
124.	6.93	153.	5.83	8.	B	37.	C
125.	6.61	154.	16.5	9.	B	38.	9x + 18
126.	48.5	155.	15.9	10.	A	39.	C
127.	47.2	156.	15	11.	B	40.	5 + 2n
128.	40	157.	17	12.	D	41.	A
129.	53	158.	2	13.	B	42.	-12.0744 - 19.5x
130.	13	159.	2.02	14.	2 - 4b	43.	A
131.	6.7	160.	1.92	15.	20n	44.	50x + 72
132.	6.35			16.	D	45.	B
133.	15			17.	C	46.	A
134.	14.73			18.	C	47.	D
135.	13			19.	25p - 8	48.	C
136.	16			20.	B	49.	B
137.	3			21.	-24b	50.	A
138.	1.79			22.	-41x + 58	51.	A
139.	1.71			23.	71p + 53	52.	-4
140.	7			24.	A	53.	7
141.	8.67			25.	-2.6n	54.	-6
142.	5			26.	A	55.	B
143.	11.5			27.	A	56.	B
144.	6.5			28.	A	57.	B
145.	5.34			29.	$\frac{25}{12} n - \frac{7}{6}$	58.	A

www.math-knots.com | www.a4ace.com

Algebra 1

Week 9		Week 10		Week 10		Week 10	
59.	4.8n	1.	A	30.	C	59.	B
60.	B	2.	A	31.	D	60.	D
61.	B	3.	A	32.	D	61.	B
62.	C	4.	A	33.	B	62.	B
63.	A	5.	B	34.	D	63.	D
64.	A	6.	C	35.	B	64.	C
65.	m + 2.1	7.	C	36.	D	65.	D
66.	B	8.	C	37.	B	66.	D
67.	B	9.	B	38.	C	67.	A
68.	C	10.	C	39.	D	68.	D
69.	C	11.	A	40.	A	69.	C
70.	D	12.	A	41.	C	70.	A
71.	A	13.	D	42.	C	71.	B
72.	D	14.	D	43.	B	72.	A
73.	C	15.	A	44.	A	73.	C
74.	B	16.	A	45.	B	74.	C
75.	-75x - 6	17.	B	46.	D	75.	D
76.	C	18.	B	47.	C	76.	D
77.	$\frac{11}{20}$ p - 2	19.	B	48.	A	77.	C
78.	A	20.	D	49.	A	78.	D
79.	-3 + 30x	21.	A	50.	C	79.	A
80.	B	22.	B	51.	A	80.	B
81.	A	23.	C	52.	C	81.	D
82.	A	24.	C	53.	C	82.	B
		25.	D	54.	A	83.	B
		26.	D	55.	D	84.	B
		27.	A	56.	A	85.	C
		28.	B	57.	D	86.	A
		29.	A	58.	C	87.	D

www.math-knots.com | www.a4ace.com

Algebra 1

Week 10		Week 11		Week 11		Week 11		Week 11	
88.	A	1.	B	30.	B	59.	B	88.	C
89.	A	2.	B	31.	A	60.	D	89.	A
90.	B	3.	B	32.	C	61.	A	90.	A
91.	D	4.	A	33.	B	62.	C	91.	D
92.	A	5.	B	34.	A	63.	B	92.	C
		6.	A	35.	C	64.	A		
		7.	B	36.	D	65.	A		
		8.	A	37.	A	66.	D		
		9.	A	38.	C	67.	D		
		10.	B	39.	B	68.	A		
		11.	B	40.	A	69.	A		
		12.	A	41.	A	70.	A		
		13.	C	42.	D	71.	B		
		14.	C	43.	A	72.	A		
		15.	D	44.	D	73.	B		
		16.	B	45.	A	74.	B		
		17.	A	46.	B	75.	A		
		18.	A	47.	D	76.	B		
		19.	C	48.	C	77.	B		
		20.	D	49.	B	78.	C		
		21.	B	50.	B	79.	D		
		22.	B	51.	A	80.	B		
		23.	A	52.	A	81.	B		
		24.	D	53.	B	82.	C		
		25.	B	54.	C	83.	C		
		26.	B	55.	A	84.	A		
		27.	D	56.	D	85.	A		
		28.	D	57.	B	86.	B		
		29.	D	58.	D	87.	D		

364   www.math-knots.com | www.a4ace.com

Week 12

1) D) x > 10 or x ≤ 8 :

2) D) −2 ≤ a < 10 :

3) D) −5 < m < 7 :

4) A) x ≤ −7 or x > 1 :

5) A) n > −9 or n ≤ −10 :

6) C) 1 ≤ x ≤ 3 :

7) D) 0 ≤ b < 3 :

8) D) r > 14 or r < 3 :

9) C) n ≤ −6 or n > 5 :

10) C) v ≥ 0 or v < −7 :

11) C) −14 < k ≤ −5 :

12) B) x ≤ −6 or x > 8 :

Week 12

13) B) $-3 < x < 13$:

14) D) $3 \le k \le 6$:

15) A) $m < 2$ or $m \ge 7$:

16) C) $x \ge -2$ or $x \le -5$:

17) D) $k \ge -1$ or $k \le -14$:

18) C) $n \ge 8$ or $n < -11$:

19) A) $0 < n < 4$:

20) C) $-12 \le a < 6$:

21) B) $-14 \le x < 10$:

22) C) $-1 < x < 3$:

23) A) $-12 \le r \le 7$:

24) D) $x < -1$ or $x \ge 0$:

     www.math-knots.com | www.a4ace.com

Week 12

25) A) $k \geq 0$ or $k < -8$:

26) D) $-14 < n < 5$:

27) C) $n \geq -1$ or $n \leq -4$:

28) D) $0 < x \leq 12$:

29) C) $-13 < m \leq -11$:

30) B) $-14 < n < -10$:

Algebra 1

Week 12		Week 12		Week 12		Week 12		Week 13	
31.	D	59.	B	88.	D	117.	B	1.	A
32.	C	60.	B	89.	C	118.	B	2.	D
33.	B	61.	D	90.	A	119.	B	3.	D
34.	D	62.	D	91.	C	120.	C	4.	A
35.	B	63.	B	92.	A	121.	C	5.	D
36.	D	64.	A	93.	A	122.	C	6.	C
37.	C	65.	B	94.	A	123.	D	7.	B
38.	C	66.	A	95.	D	124.	D	8.	D
39.	C	67.	D	96.	C	125.	D	9.	B
40.	C	68.	A	97.	D	126.	C	10.	D
41.	D	69.	A	98.	D	127.	C	11.	A
42.	A	70.	D	99.	C	128.	D	12.	A
43.	A	71.	B	100.	D	129.	C	13.	D
44.	A	72.	D	101.	C	130.	A	14.	D
45.	A	73.	C	102.	A	131.	A	15.	B
46.	A	74.	D	103.	D	132.	A	16.	C
47.	D	75.	D	104.	D	133.	B	17.	C
48.	D	76.	D	105.	B	134.	A	18.	A
49.	C	77.	D	106.	C	135.	D	19.	D
50.	B	78.	D	107.	A	136.	A	20.	A
51.	A	79.	B	108.	C	137.	A	21.	D
52.	C	80.	A	109.	A	138.	A	22.	C
53.	C	81.	A	110.	C	139.	B	23.	C
54.	B	82.	B	111.	A	140.	A	24.	A
55.	A	83.	B	112.	D			25.	A
56.	A	84.	D	113.	C			26.	A
57.	C	85.	D	114.	A			27.	D
58.	B	86.	C	115.	B			28.	B
		87.	C	116.	B			29.	A

www.math-knots.com | www.a4ace.com

Week 13			Week 13			Week 13	
30.	C		59.	D		88.	a < -10 or a > 10
31.	C		60.	A		89.	x ≥ 13 or x ≤ 5
32.	D		61.	A		90.	x > 5 or x < -3
33.	A		62.	B		91.	r ≥ 6 or r ≤ -8
34.	A		63.	B		92.	$-\frac{4}{3} < r < \frac{4}{3}$
35.	C		64.	D		93.	$x \geq \frac{7}{5}$ or $x \leq -\frac{7}{5}$
36.	D		65.	n ≥ 8 or n ≤ -22		94.	-21 < x < 1
37.	C		66.	x ≥ -7 or x ≤ -9		95.	x > 5 or x < 3
38.	B		67.	-6 < n < 6		96.	n > 7 or n ≤ -7
39.	B		68.	-40 ≤ k ≤ 40		97.	-8 ≤ m ≤ 8
40.	D		69.	-20 < x < 20		98.	-6 < k < 6
41.	A		70.	x > 13 or x < 1		99.	-8 < x < 10
42.	A		71.	p ≥ 1 or p ≤ 13		100.	1 ≤ x ≤ 9
43.	D		72.	x > 18 or x < -18		101.	v > 7 or v < -7
44.	D		73.	x ≥ 16 or x ≤ -16		102.	x > 150 or x < -3
45.	D		74.	-23 ≤ x ≤ 7		103.	-11 < x < 3
46.	C		75.	p <= -3 or p ≥ 3		104.	n ≥ 2 or n ≤ -2
47.	A		76.	k ≥ 8 or k ≤ -14		105.	-27 < p < 9
48.	D		77.	-6 ≤ v ≤ 6		106.	3 < m < 13
49.	C		78.	n ≥ 3 or n ≤ -21		107.	x ≤ -4 or x ≥ 4
50.	D		79.	a < -8 or a > 8		108.	p > 2 or p < -2
51.	D		80.	a > 3 or a < -3		109.	-5 < k < -3
52.	D		81.	x > 6 or x < -2		110.	-19 < x < 9
53.	C		82.	-3 < u < 9		111.	-3 < k < 3
54.	D		83.	-9 ≤ x ≤ 9		112.	x ≥ 6 or x ≤ -8
55.	C		84.	-9 < a < 9		113.	-1 ≤ k ≤ 1
56.	A		85.	-10 < n < -8			
57.	A		86.	b > 9 or b < -9			
58.	C		87.	-15 < r < 1			

       www.math-knots.com | www.a4ace.com

Week 13

114. $x \geq 6$ or $x \leq -6$
115. $-7 \leq x \leq 15$
116. $-5 < x < 5$
117. $v \geq 10$ or $v \leq -10$
118. $a \geq 7$ or $a \leq -3$
119. $-9 < v < 9$
120. $-5 < x < 5$
121. $-3 \leq n \leq 19$
122. $m > 5$ or $m < -5$
123. $x > 16$ or $x < 0$
124. $n \geq 9$ or $n \leq -9$
125. $x \geq 5$ or $x \leq -5$
126. $k > 5$ or $k < -9$
127. $-2 \leq x \leq 2$
128. $n \geq 8$ or $n \leq -4$
129. $-4 < v < 4$
130. $-3 \leq b \leq 3$
131. $b \geq 2$ or $b \leq -2$
132. $-\dfrac{17}{5} < p < 7$
133. $-8 \leq k \leq \dfrac{80}{5}$
134. $n \leq -2$ or $n \geq 8$
135. $-2 \leq n \leq 3$
136. $n < -2$ or $n > \dfrac{8}{3}$
137. $-4 \leq n \leq 10$
138. $-\dfrac{31}{9} \leq k \leq 3$

Week 13

139. $-4 < a < 3$
140. $-5 \leq b \leq 10$
141. $a \geq 1$ or $a \leq -19$
142. $n \geq 4$ or $n \leq \dfrac{21}{4}$
143. $n \geq 5$ or $n \leq -3$
144. $-\dfrac{7}{2} < m < 4$
145. $-\dfrac{8}{3} \leq r \leq 4$
146. $n \geq 4$ or $n \leq 0$
147. $-16 \leq x \leq 10$
148. $n \geq -2$ or $n \leq -8$
149. $n \leq -3$ or $n \geq \dfrac{11}{2}$
150. $-2 \leq m \leq 4$
151. $-6 \leq x \leq \dfrac{34}{9}$
152. $-10 < n < \dfrac{54}{5}$
153. $b < 10$ or $b > 4$
154. $-7 \leq n \leq \dfrac{71}{9}$
155. $-30 \leq v \leq 10$
156. $a > 3$ or $a < -1$
157. $a \geq \dfrac{6}{7}$ or $a \leq -\dfrac{18}{7}$
158. $-3 < k < 5$
159. $-6 < x < \dfrac{30}{7}$

www.math-knots.com | www.a4ace.com

Week 13

160.	$a > 1$ or $a < -\dfrac{5}{3}$
161.	$n \geq 5$ or $n \leq -\dfrac{13}{3}$
162.	$x < -1$ or $x > -\dfrac{1}{4}$
163.	$-\dfrac{61}{9} \leq a \leq 9$
164.	$p \geq -\dfrac{5}{7}$ or $p \leq -1$
165.	$n > 5$ or $n < -3$
166.	$b > -5$ or $b < -7$
167.	$x \geq \dfrac{1}{7}$ or $x \leq -\dfrac{11}{7}$
168.	$n \leq -8$ or $n \geq -6$
169.	$-6 \leq v \leq 15$
170.	$-5 \leq x \leq 1$
171.	$k < 1$ or $k > 2$
172.	$-\dfrac{9}{7} < m < 1$
173.	$\dfrac{1}{2} \leq k \leq 4$
174.	$-8 \leq x \leq \dfrac{13}{2}$
175.	$-\dfrac{37}{2} \leq n \leq \dfrac{33}{2}$
176.	$-\dfrac{49}{9} \leq x \leq 5$
177.	$-\dfrac{4}{5} < n < \dfrac{3}{5}$
178.	$-\dfrac{7}{2} \leq k \leq \dfrac{1}{2}$
179.	$-\dfrac{20}{7} \leq k \leq \dfrac{30}{7}$

Week 13

180.	$-7 \leq p \leq 3$
181.	$-1 < x < \dfrac{1}{3}$
182.	$x < -18$ or $x > 0$
183.	$x \geq \dfrac{41}{9}$ or $x < -5$
184.	$-2 \leq x \leq \dfrac{2}{9}$
185.	$-1 < a < 2$
186.	$-2 < n < 4$
187.	$b > 1$ or $b < -\dfrac{21}{5}$
188.	$b \geq \dfrac{14}{9}$ or $b < -2$
189.	$-\dfrac{24}{7} \leq b \leq 2$
190.	$n \geq \dfrac{27}{5}$ or $n \leq -7$
191.	$n > \dfrac{8}{3}$ or $n < -5$
192.	$a > \dfrac{9}{7}$ or $a < -3$
193.	$-\dfrac{1}{9} < x < 1$
194.	$-\dfrac{8}{3} < p < 0$
195.	$n \geq 6$ or $n \leq -22$
196.	$-\dfrac{15}{2} < n < 8$
197.	$-1 < n < 0$

Week 14

1.	C
2.	A
3.	C
4.	A
5.	C
6.	D
7.	C
8.	C
9.	D
10.	B
11.	A
12.	A
13.	B
14.	B
15.	A
16.	D
17.	A
18.	D
19.	B
20.	A
21.	A
22.	D
23.	D
24.	D
25.	A
26.	C
27.	B
28.	B
29.	D

 **Algebra 1**

Week 14	Week 14	Week 14	Week 14	Week 14
30. B	59. D	88. C	117. A	146. B
31. B	60. A	89. C	118. A	147. C
32. A	61. B	90. C	119. A	148. B
33. B	62. B	91. C	120. B	149. C
34. B	63. B	92. D	121. B	150. B
35. D	64. C	93. B	122. D	151. A
36. A	65. D	94. D	123. C	152. A
37. A	66. C	95. D	124. B	153. A
38. A	67. C	96. C	125. B	154. B
39. C	68. B	97. A	126. B	155. C
40. C	69. C	98. B	127. C	156. C
41. B	70. A	99. B	128. C	
42. D	71. D	100. D	129. A	
43. B	72. A	101. B	130. C	
44. C	73. D	102. C	131. B	
45. A	74. D	103. A	132. C	
46. A	75. D	104. B	133. B	
47. C	76. B	105. B	134. D	
48. B	77. B	106. A	135. D	
49. A	78. C	107. D	136. B	
50. D	79. C	108. A	137. D	
51. D	80. C	109. C	138. A	
52. C	81. D	110. D	139. A	
53. B	82. A	111. C	140. A	
54. C	83. C	112. D	141. C	
55. D	84. B	113. A	142. A	
56. C	85. C	114. A	143 A	
57. A	86. C	115. D	144. B	
58. A	87. C	116. A	145. A	

www.math-knots.com | www.a4ace.com

Week 15

1.　Domain : { − 5 , 0 , 3, 6, 7 }

　　Range : { 0 , 6 , 30 , 42}

2.　Domain : { − 8 , -6 , -1, 0, 10 , 11 }

　　Range : { -33 , -25 , -5 , -1 , 39 , 43 }

3.　Domain : { − 4 , − 3 , − 2, 1, 5 }

　　Range : { 7 , 1 , -3 , 25}

4.　Domain : { − 4 , − 2 , − 1, 1, 2, 5 }

　　Range : { -4 , -6 , -1 , 110}

5.　Domain : { − 9 , − 7 , − 2, 1, 6, 10 }

　　Range : {-32 , -24 , -4 , 8 , 28 , 44}

6.　Domain : { − 12 , -15 , -5, 10, 20, 11}

　　Range : { -44 , -56 , -16 , 44 , 84 , 48}

7.　Domain : { − 8 , − 4 , − 1, 2, 4, 10 }

　　Range : { 66 , 18 , 3 , 6 , 102}

8.　Domain : { − 11 , -10 , -5, 10, 12, 13 }

　　Range : { 122 , 101 , 26 , 145 , 170}

9.　Domain : { − 18 , -15 , -11, 12, 17, 80 }

　　Range : { -20 , -17 , -13 , 10 , 15 , 78}

10.　Domain : { − 25 , − 33 , − 45, 50, 64, 77 }

　　Range : { 29 , 37 , 49 , -46 , -60 , -73}

11.　Domain : { − 4 , − 3 , − 2, 1, 2 , 5 }

　　Range : { -52 , -18 , -2 , 2 , 110 }

12.　Domain : { − 4 , − 2 , − 1, 1, 2, 5 }

　　Range : { -55 , -5 , 2 , 0 , 5 , 116 }

13.　Domain : { − 11 , -14 , -2, 1, 7, 10 }

　　Range : { -30 , -39 , -3 , 6 , 24 , 33 }

14.　Domain : { − 12 , -15 , -5, 10, 20, 11 }

　　Range : {-41 , -50 , -20 , 25 , 55 , 28 }

　　　　　　www.math-knots.com | www.a4ace.com

Week 15

15.   Domain : { – 8 , – 4 , – 1 , 2 , 4 }

      Range : { -59 , 4 , 13 , 69 }

16.   Domain : { – 4 , – 10 , – 5, 1, 2, 3 }

      Range : { -64 , -1000 , -125 , 1 , 8 , 27}

17.   Domain : { – 9 , – 7 , – 2 , 1, 6, 10 }

      Range : { -16 , -12 , -2 , 4 , 14 , 22 }

18.   Domain : { – 12 , -15 , -5, 10, 20, 11 }

      Range : { -21 , -27 , -7 , 23 , 43 , 25 }

19.   Domain : { – 5 , – 3 , – 2, 1, 2, 5 }

      Range : { -175 , -45 , -6 , -1 , 0 , 75 }

20.   Domain : { – 5 , – 2 , – 1, 1, 2, 5 }

      Range : { 27 , 6 , 3 }

21.   Domain : { – 4 , – 3 , – 2 , 1 , 2 , 5 }

      Range : { 0 , -15 , 4 , 5 , 24 , 225 }

22.   Domain : { – 4 , – 2 , – 1, 1, 2, 5 }

      Range : { -8 , -2 , 1 , 7 , 10 , 19 }

23.   Domain : { – 4 , – 2 , 0, 1, 2 , 3 }

      Range : { 36 , 8 , -4 , 0 }

24.   Domain : { – 4 , – 2 , – 1, 1, 2, 5 }

      Range : { 0 , 2 , 3 , 5 , 6 , 9 }

25.   Domain : { – 4 , – 3 , – 2, 1, 2, 5 }

      Range : { 11 , 4 , -1 , -4 , 20 }

26.   Domain : { – 4 , – 2 , – 1 , 1 , 2 , 5 }

      Range : { -3 , 1 , 3 , 7 , 9 , 15 }

27.   Domain : { – 5 , – 1 , – 2, 1, 0, 5 }

      Range : { 26 , 2 , 5 , 1 }

28.   Domain : { – 3 , – 2 , – 1, 1, 6, 7 }

      Range : { 4 , -1 , -4 , 31 , 44 }

          www.math-knots.com | www.a4ace.com

Week 15

29.	Domain : { − 11 , -14 , -2, 1, 7, 10 }
	Range : { -22 , -28 , -4 , 2 , 14 , 20 }
30.	Domain : { − 12 , -15 , -5, 10, 20, 11 }
	Range : {148 , 229 , 29 , 104 , 404 , 125 }
31.	A
32.	D
33.	C
34.	D
35.	A
36.	C
37.	A
38.	D
39.	B
40.	B
41.	B
42.	A
43.	C
44.	C
45.	C
46.	A
47.	B
48.	C
49.	C
50.	B
51.	A
52.	C
53.	D
54.	D
55.	C

Week 15

56.	C
57.	C
58.	C
59.	A
60.	A
61.	A
62.	A
63.	C
64.	D
65.	D
66.	D
67.	C
68.	B
69.	B
70.	A
71.	D
72.	A
73.	B
74.	D
75.	A
76.	C
77.	A
78.	D
79.	A
80.	D
81.	A
82.	D
83.	D
84.	D

www.math-knots.com | www.a4ace.com

Week 15		Week 15		Week 15		Week 15	
85.	B	114.	C	143.	B	172.	C
86.	B	115.	C	144.	D	173.	D
87.	D	116.	C	145.	B	174.	A
88.	A	117.	D	146.	D	175.	D
89.	C	118.	C	147.	B	176.	D
90.	A	119.	A	148.	B	177.	B
91.	B	120.	C	149.	A	178.	D
92.	A	121.	D	150.	C	179.	B
93.	B	122.	C	151.	D	180.	9
94.	A	123.	D	152.	D	181.	A
95.	B	124.	A	153.	B	182.	A
96.	C	125.	A	154.	B	183.	B
97.	B	126.	A	155.	B	184.	A
98.	A	127.	A	156.	C	185.	6
99.	B	128.	A	157.	B	186.	1
100.	A	129.	A	158.	B	187.	5
101.	A	130.	C	159.	D	188.	-9
102.	A	131.	D	160.	D	189.	-1
103.	D	132.	B	161.	A	190.	-2
104.	D	133.	D	162.	A	191.	6
105.	B	134.	D	163.	D	192.	4
106.	C	135.	B	164.	D	193.	2
107.	B	136.	C	165.	B	194.	3
108.	B	137.	D	166.	A	195.	5
109.	D	138.	C	167.	D	196.	3
110.	D	139.	C	168.	C	197.	316
111.	B	140.	B	169.	A	198.	-175
112.	C	141.	D	170.	B	199.	275
113.	A	142.	D	171.	C	200.	249

376   www.math-knots.com | www.a4ace.com

Week 15		Week 16		Week 16	
201.	-3123	1.	D	30.	A
202.	859	2.	x ≥ 9 or x ≤ 3	31.	C
203.	186	3.	A	32.	B
204.	-204	4.	A	33.	A
205.	-794	5.	B	34.	A
206.	117 , 137	6.	A	35.	D
207.	-44 , -54	7.	C	36.	C
208.	795 . 995	8.	C	37.	C
209.	-43 , -52	9.	x < -7 or x > 9	38.	$-\dfrac{15}{7} < r < 1$
210.	-10 , -16	10.	D	39.	D
211.	71 , 79	11.	v ≥ 11 or v ≤ 9	40.	C
212.	34 , 38	12.	C	41.	1 < a < 17
		13.	D	42.	B
		14.	C	43.	D
		15.	C	44.	C
		16.	x ≥ 1 or x ≤ -1/5	45.	A
		17.	B	46.	2 ≤ n ≤ 3
		18.	A	47.	D
		19.	B	48.	A
		20.	$-\dfrac{24}{7} \le X \le 4$	49.	$-\dfrac{17}{4} \le n \le 4$
		21.	D	50.	p ≥ 26 or p ≤ -8
		22.	D	51.	D
		23.	A	52.	b ≥ -1 or b ≤ -3
		24.	C	53.	-40 < r < 10
		25.	r ≥ 9 or r ≤ -9	54.	A
		26.	B	55.	-1 ≤ x ≤ 1
		27.	D		
		28.	C		
		29.	D		

 www.math-knots.com | www.a4ace.com

Week 16

56.	A
57.	A
58.	B
59.	A
60.	B
61.	Domain : { -7, -3 , -2 , 1 , 2 , 7 }
	Range : { -27, -11, -7, 5 , 9 , 29 }
62.	A
63.	C
64.	$r > \dfrac{11}{7}$ or r < 1
65.	A
66.	B
67.	Domain : { -5, -3, -2, 1, 2, 5 }
	Range : { 5, -3, -4, 12, 45 }
68.	A
69.	B
70.	m ≤ -4 or m ≥ 7
71.	B
72.	A
73.	B
74.	-18 ≤ v ≤ 10
75.	A
76.	Domain : { -6, -5, -1, 7, 10 }
	Range : { 54 , 40 , 4 , 28 , 70 }
77.	B
78.	C

Week 16

79.	116 , 146
80.	-69
81.	B
82.	-7
83.	-9
84.	811 , 1011
85.	C
86.	-40 , -45

Week 17

1.	C
2.	D
3.	B
4.	D
5.	D
6.	C
7.	B
8.	A
9.	D
10.	C
11.	B
12.	A
13.	B
14.	D
15.	D
16.	C
17.	B
18.	D
19.	D
20.	A
21.	B
22.	C
23.	B
24.	C
25.	A
26.	C
27.	D
28.	A
29.	C

Week 17		Week 17		Week 17		Week 17	
30.	A	59.	D	88.	D	113.	$\frac{1}{4}$
31.	A	60.	C	89.	C	114.	$-\frac{3}{2}$
32.	D	61.	D	90.	B	115.	Undefined
33.	C	62.	B	91.	A	116.	$\frac{3}{4}$
34.	D	63.	B	92.	B	117.	5
35.	D	64.	C	93.	C	118.	$-\frac{3}{2}$
36.	D	65.	D	94.	A	119.	-1
37.	C	66.	B	95.	C	120.	$\frac{3}{2}$
38.	D	67.	B	96.	D	121.	-5
39.	D	68.	D	97.	B	122.	-2
40.	C	69.	B	98.	B	123.	5
41.	A	70.	D	99.	B	124.	-5
42.	B	71.	D	100.	D	125.	1
43.	B	72.	C	101.	-2	126.	0
44.	B	73.	C	102.	-1	127.	-3
45.	D	74.	B	103.	$-\frac{4}{3}$	128.	2
46.	B	75.	A	104.	-2	129.	-2
47.	C	76.	D	105.	$\frac{1}{5}$	130.	-2/3
48.	C	77.	D	106.	3	131.	$-\frac{21}{11}$
49.	B	78.	B	107.	-2	132.	5
50.	D	79.	C	108.	3	133.	$\frac{4}{3}$
51.	B	80.	C	109.	$\frac{1}{5}$	134.	x-intercept = 4.2, -0.2
52.	A	81.	D	110.	$\frac{6}{5}$		y-intercept = -1
53.	B	82.	D	111.	$\frac{1}{2}$		
54.	D	83.	D	112.	$-\frac{4}{5}$		
55.	D	84.	C				
56.	C	85.	A				
57.	B	86.	B				
58.	B	87.	D				

www.math-knots.com | www.a4ace.com

Algebra 1

Week 17

135.	$-\dfrac{36}{5}$
136.	$-\dfrac{7}{2}$
137.	$\dfrac{10}{7}$
138.	x-intercept = 2.8 , 7.2 y-intercept = -20
139.	x-intercept = 3 , 5 y-intercept = -3
140.	y-intercept = 13 x-intercept = -2.2 , -5.8
141.	y-intercept = 4 x-intercept = 0.8 , 5.1
142.	y-intercept = -8 x-intercept = 2.5 , 7.5
143.	y-intercept = -3 x-intercept = -3 , -1
144.	y-intercept = 0 x-intercept = 0
145.	y-intercept = -12 x-intercept = -2 , -6
146.	y-intercept = 3 x-intercept = $\dfrac{3}{8}$
147.	y-intercept = 3 x-intercept = -5
148.	y-intercept = -2 x-intercept = $-\dfrac{10}{3}$

Week 18

1.	C
2.	A
3.	C
4.	B
5.	B
6.	A
7.	C
8.	D
9.	D
10.	A
11.	D
12.	A
13.	C
14.	B
15.	C
16.	A
17.	A
18.	B
19.	A
20.	A
21.	D
22.	C
23.	B
24.	B
25.	B
26.	B
27.	C
28.	C
29.	C

Week 18

30.	B
31.	C
32.	B
33.	D
34.	B
35.	A
36.	C
37.	D
38.	C
39.	D
40.	A
41.	B
42.	A
43.	D
44.	D
45.	B
46.	C
47.	B
48.	B
49.	A
50.	D
51.	D
52.	A
53.	A
54.	C
55.	D
56.	D
57.	A
58.	A

Week 18

59.	B
60.	B
61.	D
62.	D
63.	C
64.	B
65.	D
66.	C
67.	A
68.	D
69.	A
70.	C
71.	C
72.	A
73.	A
74.	D
75.	A
76.	B
77.	D
78.	C
79.	B
80.	D
81.	C
82.	B
83.	C
84.	A
85.	C
86.	B
87.	A

www.math-knots.com | www.a4ace.com

Algebra 1

Week 18	Week 18	Week 18	Week 18	Week 18
88. D	117. A	146. C	175. D	204. A
89. B	118. B	147. B	176. B	205. A
90. B	119. A	148. D	177. A	206. B
91. B	120. B	149. C	178. A	207. D
92. B	121. D	150. B	179. A	208. A
93. D	122. D	151. D	180. D	209. C
94. C	123. B	152. B	181. B	210. C
95. C	124. A	153. A	182. A	211. D
96. C	125. C	154. A	183. B	212. B
97. D	126. A	155. C	184. B	213. B
98. D	127. A	156. C	185. D	214. D
99. C	128. B	157. A	186. C	215. A
100. C	129. A	158. A	187. D	216. B
101. C	130. D	159. A	188. D	217. B
102. D	131. C	160. C	189. C	218. C
103. D	132. C	161. B	190. A	219. B
104. B	133. C	162. C	191. D	220. B
105. B	134. C	163. B	192. D	
106. C	135. D	164. C	193. A	
107. A	136. C	165. D	194. A	
108. D	137. D	166. C	195. B	
109. C	138. B	167. A	196. A	
110. B	139. A	168. D	197. A	
111. D	140. C	169. A	198. A	
112. D	141. C	170. C	199. C	
113. C	142. B	171. D	200. B	
114. A	143. C	172. C	201. C	
115. A	144. C	173. C	202. B	
116. C	145. D	174. C	203. A	

Made in the USA
Middletown, DE
04 April 2025

73771291R00212